ORDINARY STATISTICS

by

JAMES A. WALKER M.A. M.Ed

Head of Maths Department, Dollar Academy

and

MARGARET M. McLEAN B.Sc

Assistant Maths Mistress, Stranraer Academy

EDWARD ARNOLD

First published 1973
by Edward Arnold [Publishers] Ltd
25 Hill Street
London W1X 8LL

Reprinted 1975
ISBN 0 7131 1739 7

Filmset by Keyspools Ltd, Golborne, Lancs
Printed by Tinling (1973) Limited
Prescot, Merseyside
(a member of the Oxley Printing Group Ltd)

TO THE TEACHER

Statistics in the form of graphs, figures, opinion polls etc. are playing an increasingly large part in our everyday life. It is most important that children leaving school should have some knowledge of how this information is collected, what the various types of graph mean and above all how reliable the information is likely to be. A critical and perhaps even cynical view of the data presented by the popular press, T.V., advertisements and so on, should be fostered.

The purpose of this book is to introduce the basic ideas of Statistics and Probability in a modern way. It is intended to be a complete course in Statistics from first year up to O-Level. The subject matter covers both the English Associated Examining Board syllabus and the Scottish syllabus. These are far from being identical and so teachers should omit the sections or chapters not required for their own students; this may be done without any loss of continuity. For example, Scottish students do not require to cover the chapters on Scatter Diagrams or Regression Lines etc.; and English students need not study the chapters on the Binomial Distribution, Significance Tests etc.

Some rather complicated mathematics has been included for any Sixth Form students who may be interested to pursue the subject more deeply. Students following a purely O-Level course would be well advised to omit the following sections which are not essential to the development of the subject— calculation of standard deviation from an assumed mean, Permutations and Combinations, the proof of the Binomial Theorem.

Unfortunately, in some parts of the work there is bound to be some sheer, arithmetical "slog". This is unavoidable by the very nature of the subject. To minimise this as much as possible, the use of such aids as calculating machines, slide rules, mathematical tables for square roots etc. should be encouraged, provided they are available, of course.

As much as possible, the students should collect their own data, decide how to present it graphically and how to interpret what they have found out.

Three figure mathematical tables have been used for all the calculations.

In Statistics textbooks, we find a number of different notations used for the arithmetic mean, standard deviation etc. The notation and formulae used in this book are shown on page iv.

NOTATION AND FORMULAE

$N, X, \overline{X}, s$ refer to samples and n, x, μ, σ to populations

Arithmetic Mean:

$$\overline{X} = \frac{\Sigma X}{N} \quad \text{or} \quad \frac{\Sigma f X}{\Sigma f}$$

$$\mu = \frac{\Sigma x}{n} \quad \text{or} \quad \frac{\Sigma f x}{\Sigma f}$$

Standard Deviation:

$$s = \sqrt{\frac{\Sigma(X - \overline{X})^2}{N}} \quad \text{or} \quad \sqrt{\frac{\Sigma f(X - \overline{X})^2}{\Sigma f}}$$

$$\sigma = \sqrt{\frac{\Sigma(x - \mu)^2}{n}} \quad \text{or} \quad \sqrt{\frac{\Sigma f(x - \mu)^2}{\Sigma f}}$$

Standard Variable:

$$z = \frac{X - \overline{X}}{s} \quad \text{or} \quad \frac{x - \mu}{\sigma}$$

Binomial Distribution:

$$\mu = Np; \quad \sigma = \sqrt{(Npq)}$$

Samples of Size N (N large) from an infinite population:

$$\mu_{\overline{x}} = \mu; \quad \sigma_{\overline{x}} = \frac{\sigma}{\sqrt{N}}$$

INTRODUCTION

Statistics is the name given to the science of collecting facts and studying or analysing them. The facts or "data" can cover a tremendous range of subjects. They may be quite simple facts like how many bottles of milk a family buys every week or the goal average of a favourite football player. On the other hand, they may be extremely complex facts of the world of industry or nuclear physics and which can only be analysed by using computers.

If the data available on a subject is incomplete the answers we get to problems will only *probably* be true and are sometimes called predictions. An example of this is found in weather forecasting. Information about temperatures, wind, rain, barometric pressure etc. is collected in weather stations all over the world, on land and sea and even high in the air by means of weather balloons. When the weatherman has studied all this data, he has to predict what is *likely* to happen and, of course, he knows he is taking the risk of being wrong in his forecast.

Although statistics are becoming more and more widely used these days, their use is not a modern science by any means. For centuries men have collected information of various kinds and then put it to use. The Romans, for instance, conducted accurate censuses to count the number of people in the countries they conquered. It tells us in the Bible that Jesus was born in Bethlehem because Mary and Joseph had to travel there for the Roman Census. The Romans required that, on a certain date, every man (with his family) had to return to his birthplace to be counted. The purpose of the Roman Census was mainly to count the people who should have been paying taxes.

In this country, a national census is taken every ten years. Every household in the land is given a form with certain questions on it. This form must be filled in with information about everyone in the house at a certain hour on a certain date. People who are travelling at this time, are included in the census form at their destination whether it be a household or hotel. Institutions like jails, hospitals, boarding schools and so on also complete census forms. In this way almost every single person in the country (apart from tramps out on the road) is accounted for. The questions on the census form vary, to some extent, from one census to another depending on what facts the Government wants to know. Anyone who refuses to complete a census form is taken to court and fined or jailed.

Conducting a census is a mammoth and very expensive task. It cost the Government £3 million for the 1961 census. Even the distribution of forms involves great organisation. This is done by the Registrars of Births, Deaths and Marriages of each district in the country. The Registrars sub-divide their districts into small areas and employ a responsible person, over 21 years old, to distribute the forms in each area. This person must visit each house on his list and personally hand the form to an adult explaining what it is and that it must be filled in. As you may imagine, many visits will have to be made to some houses before anyone is found at home. After the census date, the forms have to be collected in the same way. Then all the forms are gathered together and the facts from all the forms in the country have to be tabulated. It is usually a year or two after a census before the results of it are ready to be published. Obviously a census of this comprehensive nature gives the Government excellent data about the population.

Statistics are all around us today. Whenever we open a newspaper we see facts and figures about something—what the "average" man spends on drink and tobacco; unemployment figures; road accident figures; business conditions; opinion polls. Often these facts are accompanied by a graph, diagram or chart of some kind. When we turn to the sports pages of the paper, we find them full of statistics about football, cricket, golf, etc.

Advertising is full of "statistics" too: '4 out of 5 housewives can't tell margarine from butter! 9 out of 10 people prefer "Sparkle" for washing up!' The advertisers, however, never tell us how they arrived at these figures.

In science and medicine, statistics are vitally important. Doctors, researching into new medicines and new treatments for diseases, very carefully collect and study the data from their experiments. In this way, they discover how effective the medicine or treatment may be. New drugs must be exhaustively tested for several years before they are released for general use.

Manufacturers and business men spend hundreds of thousands of pounds in building laboratories and employing scientists and mathematicians to conduct research into a great variety of products like modern building materials, new types of car engines, chemicals, new synthetic fibres and materials, etc. A great deal of this research is concerned with collecting data and drawing conclusions from it. If the results are not what is wanted then a new batch of experiments is set up, more data is collected and further conclusions drawn. This process may have to be repeated for months or even years, before the scientists achieve what they are after.

As you see, Statistics is part of every aspect of our daily life and in this book we shall be looking at how data can be collected to give as accurate a picture as possible, and seeing how we can interpret this data.

CONTENTS

THE COLLECTION AND REPRESENTATION OF DATA

Collection of data

This aspect of Statistics may, at first glance, appear to present no problems but in fact one of the most difficult tasks of a statistician is collecting accurate data.

Of course some facts are very easy to collect. For instance, there can be no doubt about the number of goals scored by football teams during a season as they are recorded by the Football League. On the other hand, if you ask a housewife how many loaves of bread she had bought every week for the last month, it is very unlikely that she would be able to tell you accurately.

There are different types of data which we may wish to collect—there are 'measurable' things like heights, weights; 'countable' things like the number of children in a family, the number of deaths due to motor accidents etc., and there is the information obtained by asking people questions. Opinion polls—very often on political issues—give the results of the latter.

The difficulties arise when we start to ask people questions.

The most reliable way of asking questions is by a questionnaire. Once a questionnaire has been compiled it has to be decided whether it should be sent through the post or whether an interviewer should be used to read aloud the questions and record the answers. The second method is much better. It ensures that each question is answered, and even more important, that each question is understood. The trouble with sending questionnaires by post, to be answered by the people concerned on their own, is that often the questions are not understood and even more often the questionnaires are never filled in and returned. So if you wish to collect information by sending out questionnaires by post you can only rely on a very small percentage being returned.

Having considered how to ask the questions, we must think very carefully of *what* questions are to be asked. They must be very carefully worded so that as far as possible there can be no doubt about their meaning. For instance, if we ask: 'How many good meals did you have last week?' the person questioned has to decide for himself exactly what is meant by a good meal—does it mean a substantial three course meal or does it mean a well cooked meal? Also, people's ideas will vary about what constitutes a substantial meal or a well cooked one.

We must also recognise the fact that on being questioned people either intentionally or sub-consciously may be dishonest. If we ask a man in the presence of his wife how much he spends on gambling every week, we may receive an answer very different from one obtained from him privately! Also in a sense trying to please you or to put themselves in a good light, people often give you the answer they think is *expected*. The person being questioned should never be told what the survey is about. This also helps to avoid being given the 'right' answers.

A series of questions which builds up the required answer may be formulated. These may be self-checking in some way. For example, the same questions may be put in several ways and any inconsistency in the replies is revealed. If any inconsistency is very marked then that particular question-naire must be scrapped.

Suppose you wish to conduct a survey of how many people have been seriously ill over the past year. The bald question would be: 'Have you been seriously ill over the past year?' Various types of responses are likely to be forthcoming. Some people would be unwilling to admit, even to themselves, that they had been very ill; some people would not know how ill they had been and some would be inclined to exaggerate a minor illness. You would, of course, get a number of perfectly honest replies—but which ones and how many? In a case like this, a series of 'leading questions' can build up a correct picture for the investigator.

1. Did you attend the doctor last year?
2. Were you attending the doctor regularly?
3. Were you confined to bed for more than a day or two?
4. Did you require to go to hospital?
5. How long were you in hospital?
6. What kind of treatment did you receive —an operation or medication etc.?
7. How long a convalescence did you have?

These 'small' questions are more likely to be answered correctly, since the person being interviewed usually does not sense the drift of the questions and so gives a 'loaded' answer.

In designing a successful questionnaire, certain simple rules must be adhered to:

1. The questions must be very simply worded and straightforward.

2. They should be of such a nature that the answers are 'yes' or 'no', or a precise answer such as a number or place.

3. There must be no ambiguity—only one interpretation should be possible.

4. Any question of a personal nature should be very tactful.

Having now considered how to ask questions and what kind of questions to ask, the remaining problem is whom you ask. This is the most difficult problem to solve.

If some information is wanted about a particular small group of people, then each person in the group can be questioned individually. If, however, you want information about a large number of people, say the population of a large city or indeed the population of Great Britain, then it is impossible to question every one of them (except, of course, in a Government Census). In this situation, and this is the most usual one, we must question a *sample* of the population.

How do we choose a sample which is representative of the whole group? If you pick out every hundredth name in a tele-phone directory, you are not considering people who have no telephones. This would give a *biased* sample. If you stopped every fiftieth person in a street near a factory at five o'clock, then your sample would be biased towards factory workers. If you stopped every fiftieth person at ten thirty a.m. near a busy shopping centre, then your sample would be biased towards housewives.

You may, of course, *want* a sample which is biased towards a certain group. If a market research firm is dealing with the effect of advertising on sales of soap powder, then the sample should be biased towards the group of people who usually choose the brand of soap powder that a family buys, namely the housewives.

One way of obtaining a reasonably unbiased sample is to use the electoral roll of the town and go through it picking out every fiftieth or hundredth name (depending on how many people are to be in the sample). This, however, does not take into account people under the age of eighteen years, a factor that could be allowed for by making sure that a number of young people were also questioned.

Another kind of sample to take is a *random sample*; this is dealt with in the chapter on Sampling. There is a rather complicated way of obtaining a really *representative* sample of a given population, which is used by good market research firms. but it involves a great deal of work and is outside the scope of this book.

The purpose of choosing a sample need not be for the answering of questions, it may be for finding out other information such as weights, or heights of the members of the sample.

Representation of data

Once we have collected our data, we need to present it in an attractive and eye-catching manner. The obvious way of doing this is by means of some kind of graph or chart. These graphs can be made attractive by the use of the various types of modern colouring pens—felt tipped and fibre tipped pens etc. The colours are varied and very bright—much better than colouring pencils, but much more expensive.

We often see graphs of different kinds in newspapers, magazines etc. Start a collection of these graphs. By the time you have finished this chaper, you should be able to decide whether the graphs you have collected are good representations of the data or misleading ones.

Pictographs

One of the most arresting ways of illustrating statistics is by using a graph in the form of pictures. This kind of graph is called a **pictograph** (sometimes **pictogram**). These pictures may be of cars, houses, milk bottles, aeroplanes, people etc., reduced in size but drawn to a definite scale for purposes of comparison.

Local councils or corporations often use some kind of pictograph to show how the money they take in from the rates is used. Each council must produce a balance sheet every year to inform the ratepayers how their money has been spent. Many councils put a great deal of effort into showing this information in the form of a pictograph. The reason for this, is that the information is more easily seen and appreciated by people when presented by pictures, than in a list of facts and figures.

Let us imagine that a certain local council presented the following balance sheet (page 4) for the ratepayers one year.

Fig. 1.1

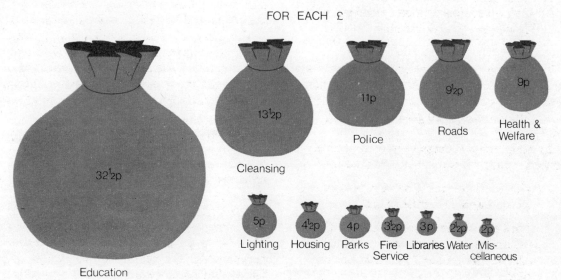

FOR EACH £

13½p — Cleansing

11p — Police

9½p — Roads

9p — Health & Welfare

32½p — Education

5p — Lighting

4½p — Housing

4p — Parks

3½p — Fire Service

3p — Libraries

2½p — Water

2p — Miscellaneous

For each £ spent:		
Education	cost	$32\frac{1}{2}$p
Cleansing	,,	$13\frac{1}{2}$p
Police	,,	11p
Roads	,,	$9\frac{1}{2}$p
Health and Welfare	,,	9p
Lighting	,,	5p
Housing	,,	$4\frac{1}{2}$p
Parks	,,	4p
Fire Service	,,	$3\frac{1}{2}$p
Libraries	,,	3p
Water	,,	$2\frac{1}{2}$p
Miscellaneous	,,	2p

Figure 1.1 (page 3) shows one way of illustrating these figures in a pictograph. The 'money bags' are drawn in proportion to the amounts of money (in respect of length).

Make up a pictograph of your own to illustrate the same figures. Then compare your pictographs.

Try to obtain pictographs of this kind put out by local councils (try the local public library) and study them.

Figure 1.2 shows the number of cases of tinned peaches sold by four shops in a month. Study the pictograph and see if you think this is an accurate way to represent the data. If you think it is not, give reasons for your answer. This pictograph reveals clearly the weakness of this kind of graph.

Shop A sold twice as many cases as Shop B and so the cube for Shop A is of side 4 cm while the cube for Shop B is of side 2 cm i.e. the dimensions for the first cube

Represents 10

Shop A   Shop B   Shop C   Shop D

Fig. 1.3

are twice what they are for the second *but* the resulting picture produces the impression of volume and Shop A has *apparently* sold a great deal more than twice what Shop B sold. How much more according to volume?

In drawing pictographs of this kind, even when keeping the linear dimensions in proportion to the data, the resulting pictures can be in a very different ratio.

Fig. 1.2

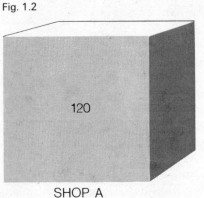

120   60   40   30

SHOP A   SHOP B   SHOP C   SHOP D

A much more accurate pictograph of the data is that shown in figure 1.3 opposite. There one small case is used to represent every ten cases.

EXERCISE A

1. Study this pictograph and answer the questions below.

Fig. 1.4 The Number of Hours of Bright Sunshine in Jan. 1965

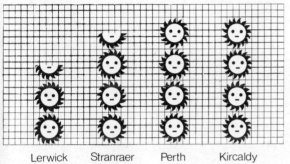

Lerwick   Stranraer   Perth   Kircaldy

 Represents 10 hours

a) Why has this been drawn on squared paper?
b) What does one picture of the 'sun' represent?
c) How many hours of bright sunshine did each of the four places have?
d) What does half a 'sun' stand for?
e) What would a quarter of a 'sun' stand for?

2. In the same year the following hours of sunshine were also recorded in January.

Stornoway—36, Cardiff—32, Lowestoft—48, Paisley—22.

Draw a pictograph like the one in question 1. This time let a 'sun' represent eight hours of sunshine.
What will half a 'sun' represent?
What will quarter of a 'sun' represent?

3. Study the following pictograph and answer the questions. (Notice how the columns of 'fish' are evenly spaced out).

Fig. 1.5 The Catches of Five Anglers in a Season's Fishing

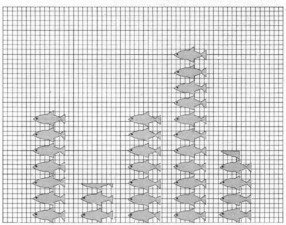

Mr.Black   Mr.Green   Mr.White   Mr.Brown   Mr.Gray

 Represents 20 fish

a) What is the scale of this pictograph?
b) Which two men had caught the same number of fish? How many did each catch?
c) Which man had a catch of 50 fish?
d) Which man had a catch of 220 fish?
e) What size of catch do you estimate Mr Gray had?
f) Which man won the fishing prize?
g) Who deserved the 'booby' prize?

4. The rainfall (in centimetres) recorded one year for these places was as follows:
Renfrew—90; Kinloss—25; Benbecula—110, Dunoon—210, Helensburgh—125.
Draw a pictograph to illustrate this using an umbrella, to represent 10 cm of rain.
Watch these points—
1. Make your umbrellas the same size.
2. Space them out evenly.
3. Have a title.
4. Show your scale.
5. Put in the names.
(This pictograph could be made very attractive, by shading the tops of the umbrellas a different colour for each town). Dunoon is a popular holiday place. Would you choose it for a holiday?

5

5. The number of bottles of milk used in a small school for each week of a month are shown below.

> 1st week — 40 bottles
> 2nd week — 36 bottles
> 3rd week — 44 bottles
> 4th week — 38 bottles

Draw a pictograph to illustrate this, using a milk bottle for your picture. This time choose your own scale. Remember to watch for the five points mentioned in question 4.

6. A small poultry farmer with free range hens sold the following numbers of eggs each month over six months. Draw a graph to show this, choosing your own scale.

> Jan. — 5 dozen   April — $14\frac{1}{2}$ dozen
> Feb. — 4 dozen   May — 12 dozen
> Mar. — 9 dozen   June — $10\frac{1}{2}$ dozen

Find out some data for yourselves and draw pictographs to show it.

Column graphs

The pictographs you were drawing and looking at in the last section are an attractive way of presenting data but they have two main drawbacks which you have probably found out for yourselves, namely,

1. They take a long time to draw well.
2. They are difficult to read accurately, when there are 'bits' of pictures.

If instead of a column of pictures, we simply draw a plain column, then we overcome both difficulties.

Figure 1.6 is a **column** graph of the number of bottles of milk used in the school in question 5 of the last exercise.

Study the graph and note the following.

1. The title tells us what the graph is about.
2. The columns are all the same thickness and are spaced out evenly along the *horizontal axis*.
3. The *vertical axis* has a scale (like a ruler) which tells us the number of bottles.
4. The columns rise to the correct height for each week measured against the vertical scale.

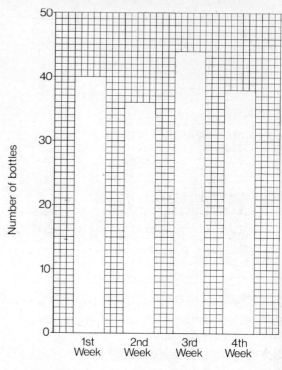

Fig. 1.6 The Number of Bottles of Milk Used in a Small School in Four Weeks

EXERCISE B

1. In a school one year, a second year class decided to investigate how all the pupils travelled to school. Figure 1.7 shows how the first year pupils travelled to school. Some of the pupils live in country areas well away from normal bus routes and these children are collected by taxis or by small vans used later in the day for transporting meals from central kitchens to the outlying primary schools.

a) What is the most usual method of travelling to school?
b) What is the least usual method of travelling to school?
c) What transport is used by 30 pupils?
d) How do 57 of the pupils travel?
e) How many cycle to and from school?
f) How many pupils are collected by taxis or meals vans?
g) How many first year pupils were present the day the survey was done?

6

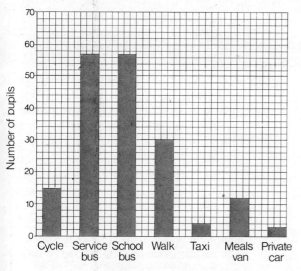

Fig. 1.7 The Means of Travelling to School of First Year Pupils

2. Conduct a similar survey either in your own class or in several classes and draw a graph of your results.

3. Study this graph and answer the questions.

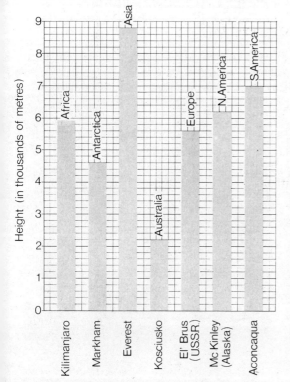

Fig. 1.8 The Highest Mountain in Each Continent

a) Which continent has the highest mountain in the world?
b) Which continent does not have any very high mountains?
c) What is the approximate height of Mount McKinley?
d) What mountain is approximately 7000 metres?
e) In which continent is there a mountain of about 5900 metres?
f) Which mountain is slightly more than half the height of Mount Everest?

4. Use your Atlas to find the heights of some of the highest mountains in Britain and draw a graph of the heights.

5. Study the graph and answer the following questions.

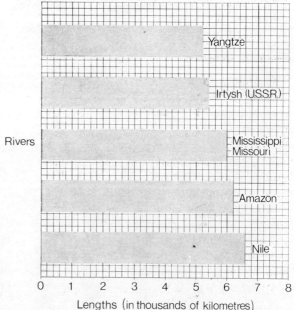

Fig. 1.9 Longest Rivers in the World

a) Why do you think the columns have been drawn horizontally for this graph?
b) What is the approximate length of the longest river in the world?
c) Which river is about 5400 kilometres long?
d) Which is about 6200 kilometres long?
e) Which two of the rivers shown are most nearly the same length?

7

6. Use your Atlas to find the lengths of the longest British rivers and draw a graph to illustrate the lengths, drawing your columns horizontally.

7. Draw a horizontal column graph to illustrate the following stopping distances for cars. By different shading differentiate between the 'thinking distance' and 'braking distance'. These stopping distances are for private cars or small vans with good drivers, perfect brakes, tyres etc. in broad daylight with good, dry roads. Larger vehicles under the same conditions may need twice these distances. On wet roads, for all vehicles, twice the normal distances are required.

Speed	Thinking Distance	Braking Distance	Overall Stopping Distance
km/h	metres	metres	metres
32	6	6	12
48	9	13.5	22.5
64	12	24	36
80	15	37.5	52.5
96	18	54	72

8. The absences shown in the graph below were recorded in a school during a 'flu epidemic in February 1968, for two consecutive school weeks. (Notice how we can use column graphs for purposes of comparison).
   a) Which of the two classes was affected first by the 'flu outbreak?
   b) On how many days were there more than 10 pupils absent in class A?
   c) On how many days were more than 10 pupils of class B absent?
   d) What was the largest number of pupils to be absent on any day in either class?
   e) Between what two days was there the biggest jump in the absences for (i) class A (ii) class B?
   f) On what days were there more than eight pupils absent in both classes?

9. A pupil conducted a survey in two first year classes on their favourite flavours in potato crisps. The results are shown at the top of the next page. Draw a graph similar to the one in question 8 to illustrate this—choosing your own method of shading.

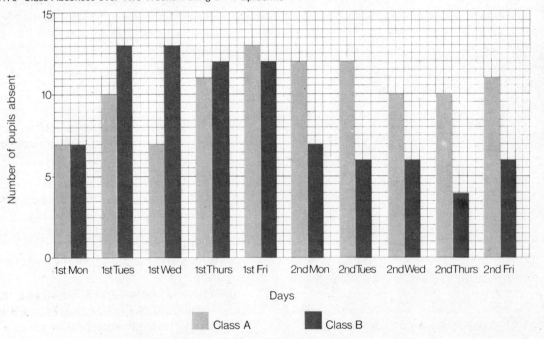

Fig. 1.10 Class Absences over Two Weeks During a Flu Epidemic

8

Crisps' Flavour	Class A	Class B
Smokey Bacon	6	14
Cheese and Onion	7	13
Plain	1	0
Salt and Vinegar	18	5

10. Draw a graph grouping the subjects for each boy together to illustrate the examination marks of these boys in the subjects shown.

Boy	English	Maths	Art
Alan	60	50	60
Bert	63	50	53
Colin	87	65	30
David	60	43	75

11. Draw a graph to show the medal positions of the top four countries after the Mexico Olympics in 1968.

Country	Gold	Silver	Bronze
United States	45	27	34
Soviet Union	29	32	30
Japan	11	7	7
Hungary	10	10	12

Collect data for yourselves in your class or in other classes and draw graphs; such things as hair colours, eye colours, shoe sizes, weights, heights, favourite pop groups, TV stars, TV programmes etc.

Line graphs

Line graphs as the name suggests show the data by means of drawing a line. This kind of graph is very good for showing upward or downward trends, and is the kind of graph most used and *misused* in newspapers, magazines and advertisements.

Example

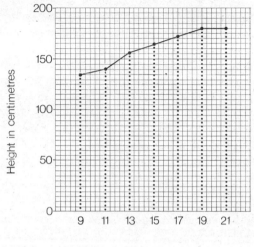

Fig. 1.11 A boy's height over several years

Notice how the points are plotted. A line is taken up from the ages to the required height. Then the points are joined with straight lines.

The boy grew most between the ages of 11 and 13 years and so the line has the steepest upward slope between those ages.

Between the ages of 19 and 21 years, he did not grow at all and so the line is level.

EXERCISE C

1. a) Does the graph on page 10 (Fig. 1.12) show what you would expect concerning the monthly number of hours of bright sunshine?

   b) What two months had less sunshine than you might expect?

   c) Between what two months was there the biggest increase in the number of hours of sunshine?

   d) Two different periods show an almost identical drop in the number of hours of sunshine. What are they?

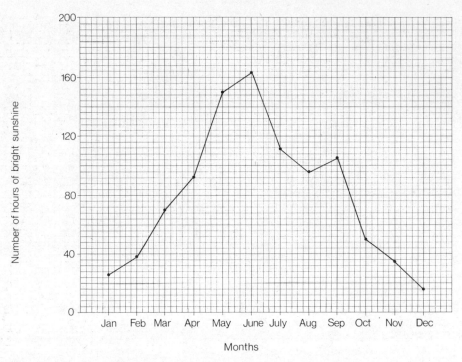

Fig. 1.12 The Number of Hours of Bright Sunshine Monthly for a Year in a town in the Far North of Britain

2. In the same year the number of hours of bright sunshine recorded in a town much further south were as follows.

Jan.	— 57 hours	July	— 188 hours
Feb.	— 84 ,,	Aug.	— 139 ,,
March	— 44 ,,	Sept.	— 146 ,,
April	— 136 ,,	Oct.	— 75 ,,
May	— 185 ,,	Nov.	— 71 ,,
June	— 209 ,,	Dec.	— 47 ,,

Draw a graph similar to the one in question 1 using these figures (or try to find out the figures for your own town) and compare and contrast the two graphs.

3. The infant mortality rates (per 1000 births) in Glasgow are shown over a period of years.

1930–34 (average)	— 102
1935–39 ,,	— 93
1940–44 ,,	— 95
1945–49 ,,	— 64
1950–54 ,,	— 37
1955–59 ,,	— 35
1960–64 ,,	— 33

Draw a line graph to illustrate this and comment on it, giving reasons for the trend it shows.

Why should the infant mortality be given as a rate and not as the actual numbers of infants who died?

4. Find out the infant mortality rates for your own town or area, draw a line graph to illustrate them and then compare and contrast the two graphs.

(The figures for questions 3 and 4 may be found in the Registrar General's Annual Reports—separate reports being published for England and Wales, and Scotland.)

5. The following graph (Fig. 1.13) shows a very different trend. Study the graph and comment on the statement that it 'represents a lesson in the value of intensive preventive medicine.'

6. The annual incidence of whooping cough in a large city from 1940–65, is as follows.

10

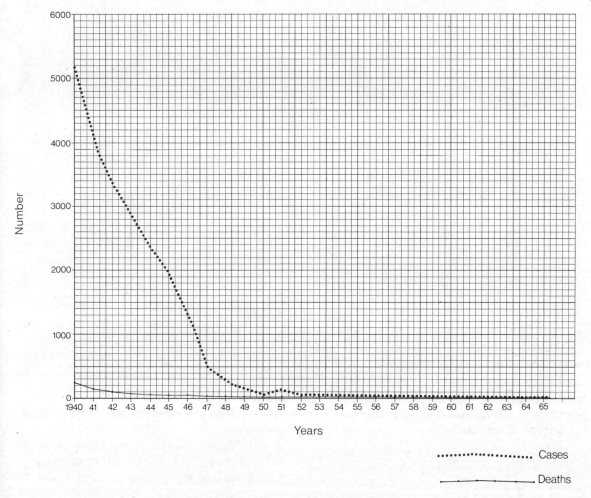

Fig. 1.13 The Incidence and Death's From Diptheria in a Large City (1940–1965)

Year		Cases	Deaths
Average	1940–44	4463	92
,,	1945–49	3321	32
,,	1950–54	4794	13
,,	1955–59	2276	3
,,	1960–64	1657	1
,,	1960	3745	4
,,	1961	824	—
,,	1962	272	—
,,	1963	2695	2
,,	1964	751	—
,,	1965	459	—

Draw line graphs to show the cases and the deaths on a *large* sheet of graph paper — being very careful of your time scale.

Comment on the trend shown in the graph.

Draw another graph to illustrate the same data as if you were trying to publicise the need for children to be immunised against whooping cough — the sort of graph you might see on a poster. Try to make up a bold, arresting title.

7. Draw a graph from the figures given at the top of page 12, to show the comparison in the annual death-rate (per 100 000 population) from pulmonary tuberculosis in Glasgow, Aberdeen and Manchester from 1955–65.

11

Comment on the trends shown and the differences between the three cities.

DEATH-RATES (PER 100 000) FROM PULMONARY TUBERCULOSIS											
Year	1955	–56	–57	–58	–59	–60	–61	–62	–63	–64	–65
Glasgow	28	25	24	26	20	19	18	18	21	14	14
Aberdeen	8	10	5	7	6	5	5	2	4	1	3
Manchester	19	15	14	10	12	12	8	11	8	8	7

8. The Registrar General of Scotland, on the basis of all the information at his disposal about death-rates, birth-rates, emigration rates etc., in 1964 estimated the population of Scotland up to 1981 as shown below.

Year	Population (in thousands)	
	Male	Female
1964	2534	2708
1969	2576	2734
1974	2650	2789
1976	2687	2817
1981	2789	2898

Draw a graph to illustrate this showing males and females separately. (Watch your time scale.)

Comment on the trend.

Try to find out the latest figures for the population (try the local registrar of births, deaths and marriages) and see whether this prediction is proving accurate. If not try to find out the reason.

Misuse of line graphs

Line graphs often appear in newspapers, magazines and advertisements and can be very uninformative and even misleading once you take a close look at them.

The 'Ten Year Growth' of a company is shown in these figures published in a daily newspaper recently.

Year (ended 31st March)	Profit before Tax (£ millions)
1971	34
1970	30.9
1969	29.8
1968	27.7
1967	25
1966	22.5
1965	21.3
1964	20.5
1963	17.8
1962	15.1

Let us look at a few ways in which these figures could be shown graphically.

Business is doing very nicely.

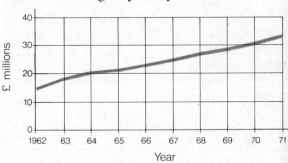

Fig. 1.14 (a)

12

Business is looking up.

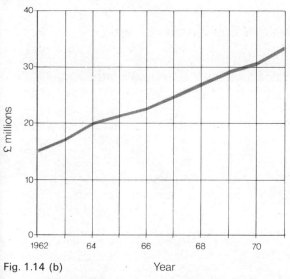

Fig. 1.14 (b) Year

Business is booming!

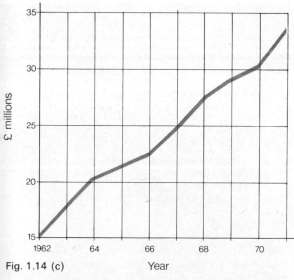

Fig. 1.14 (c) Year

As you can see, it is very easy to alter the 'look' of a line graph by adjusting one or both of the scales.

If you study newspaper graphs like the one below, you will notice that the vertical scale

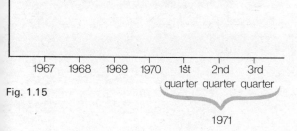

Fig. 1.15

(often numbers or amounts of money) very rarely starts at zero. The horizontal scale—usually a time scale—can be squeezed closer or spread out to give very different impressions. Indeed, sometimes the time scale is not even regular.

As you may imagine, the resulting graph was quite misleading.

Some graphs, particularly in advertisements, use very vague terms. For instance, some advertisements for headache tablets or powders show graphs of the time it takes the tablets to overcome the 'threshold of pain'—making no attempt to define this term. Still—the graph looks impressive!

We must then, when we see graphs in newspapers etc., take more than just a cursory look (which creates the impression the graphs were intended to create); we must study them carefully, looking at the scales and at the terms used. Only then can we decide what value the graph has.

Pie charts

Another way of illustrating data is by means of a **pie chart.** This is a circle divided up into sections which are usually shaded in various ways.

Figure 1.16 shows a pie chart illustrating the eye colours of a class of 36 pupils.

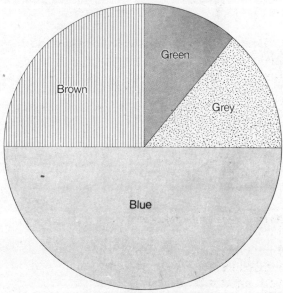

Fig. 1.16 Eye Colours in a Class of 36 Pupils

13

To read a pie chart you must first measure for each section the angle at the centre of the circle.

Thus, blue eyes has an angle of 180°
brown ,, ,, ,, ,, ,, 90°
grey ,, ,, ,, ,, ,, 50°
green ,, ,, ,, ,, ,, 40°

The whole circle represents the whole class, and a fraction of the circle represents the same fraction of the class.

The fraction of the circle representing blue eyes

$$= \frac{180}{360} = \frac{1}{2}$$

Thus, the fraction of the class with blue eyes

$$= \frac{1}{2} \text{ of } 36$$

$$= 18$$

Similarly, the number of pupils with brown eyes

$$= \frac{90}{360} \text{ of } 36$$

$$= 9$$

the number of pupils with grey eyes

$$= \frac{50}{360} \text{ of } 36$$

$$= 5$$

the number of pupils with green eyes

$$= \frac{40}{360} \text{ of } 36$$

$$= 4$$

Example

Draw a pie chart to illustrate these road deaths in Scotland one year, for boys aged up to and including 19 years.

Pedestrians	= 79
Pedal cyclists	= 5
Drivers or passengers on motor cycles	= 27
Drivers or passengers in motor cars	= 46
Total deaths	= 157

Solution

First we must change the fraction of the total deaths for each type into the angle to be used in the pie chart.

Pedestrians $\frac{79}{157}$ of total deaths.

On the pie chart, the angle required

$$= \frac{79}{157} \times 360°$$

$$= 181.2° \text{ (approx).}$$

Pedal cyclists $\frac{5}{157}$ of total deaths.

On pie chart, the angle required

$$= \frac{5}{157} \times 360°$$

$$= 11.5° \text{ (approx).}$$

Drivers or passengers on motor cycles

$$= \frac{27}{157} \text{ of total deaths}$$

On the pie chart, the angle required

$$= \frac{27}{157} \times 360°$$

$$= 61.9° \text{ (approx).}$$

Drivers or passengers in motor cars

$$= \frac{46}{157} \text{ of total deaths}$$

On the pie chart, the angle required

$$= \frac{46}{157} \times 360°$$

$$= 105.5° \text{ (approx).}$$

Fig. 1.17 Road Deaths for a Year in Scotland (Boys aged up to and including 19 Years)

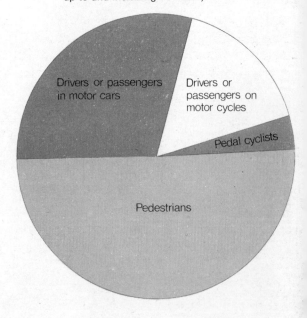

EXERCISE D

1. Measure the angles of Fig. 1.18 and calculate the number of pupils with each colour of hair.

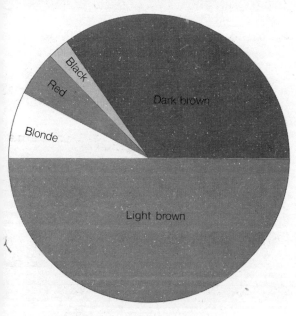

Fig. 1.18 Hair Colours in a Class of 40 Pupils

2. Draw a pie chart to illustrate the composition of the earth's atmosphere.

Nitrogen 78%
Oxygen   21%
Argon   almost 1%
(very small traces of other gases)

3. This data shows the way in which the world's catch of fish is composed of the various types of fish. Draw a pie chart to illustrate this.

33% Herring family (herrings, pilchards, sprats, sardines, shad, white bait and anchovies)
15% Demersal fish (the white fish—cod, haddock, whiting, ling, hake, and the flat fish—halibut, flounder, plaice, sole.)
31% Other Marine Species (including tuna, barracudas, mullets, perches, sharks etc.)
11% Freshwater Species
 2% Salmon
 8% Crustaceans and Molluscs (crustaceans—lobsters, shrimps etc. molluscs—oysters, mussels etc.)

Collect data of your own from your class or another class and draw pie charts to illustrate.

Disadvantages of pie charts

Pie charts have certain disadvantages.
   1. There has to be actual measurements of angles using protractors.
   2. There is quite a bit of calculation involved.
   3. The answers in most cases must be approximations.
However, pie charts have one **advantage**—they show how one whole thing is divided into parts and what size these parts are in relation to each other and to the whole.

2 ESTIMATION OF ERROR

Since much of the data that is considered in Statistics is obtained by measurement, it is important that we should understand how accurate our information is and based on this information, how accurate any further calculations will be.

It seems appropriate, then, at this stage in the course to consider how we may estimate the extent of any error that may be contained in our data.

Approximations

Significant figures

58.4819 may be written as

58.482 correct to 5 significant figures
58.48   correct to 4 significant figures
58.5   correct to 3 significant figures
58   correct to 2 significant figures
60   correct to 1 significant figure

Decimal places

58.4819 may be written as

58.482 correct to 3 decimal places
58.48   correct to 2 decimal places
58.5   correct to 1 decimal place
58   correct to the nearest whole number
60   correct to the nearest ten

EXERCISE A

1. Express the following, correct to the number of significant figures shown in the brackets.

   a) 5.632 (3)   b) 10.08 (3)   c) 10.03 (2)
   d) 0.00681 (2) e) 6.138 (1)   f) 117.2 (2)

2. Express correct to 1 decimal place.

   a) 11.66   b) 9.81   c) 69.008
   d) 0.085   e) 6.99   f) 15.849
   g) 15.850   h) 15.851

3. Express 582 733
   a) to the nearest 100 000
   b) to the nearest 10 000
   c) to the nearest   1 000
   d) to the nearest   100
   e) to the nearest   10

Absolute error

In dealing with certain data the information is found by counting and there can only be one correct answer e.g. the number of pupils on the roll of a particular school, the number of runs scored by a cricket side during a season or the number of goals scored by a hockey eleven during a winter. Other information is found by measuring, and this is a very different situation.

In measuring anything whether it be the length of a line, a period of time or the mass of an object we are limited in our accuracy by the equipment available and our own human limitations. We can never find the *exact* measure we are seeking and we must be content with an approximation. The difference between the true measure and that obtained by measurement is called the **error** (though there has been no mistake in the measuring). This error can be reduced, of course, by using more accurate instruments but a measurement can never be exact and so errors can never be eliminated completely.

It is important then that we should be aware of what error is implied by our measurements and what the maximum possible error is likely to be.

Let us consider the measurement of a straight line. Using an ordinary ruler the smallest unit of measurement that we can read off, with any accuracy is a millimetre or 0.1 cm. If our line is measured as 4.9 cm, this means that it measured 4.9 cm *to the nearest* 0.1 cm i.e. the line was *actually* between 4.85 cm and 4.95 cm long. The maximum error then is 0.05 above our nominal value and 0.05 below it. This is called the **absolute error** (Note that the absolute error is half the smallest unit of measurement. In the example above, the smallest unit of measurement was 0.1 cm and the absolute error 0.05 cm).

Example

For each of the following measurements, find
  a) the smallest unit of measurement,
  b) the absolute error,
  c) the upper and lower limits of the true measurement.
    1. 5 cm
    2. 30.6 seconds

Solution

1. a) The smallest unit of measurement is 1 cm.
   b) The absolute error is half of the smallest unit, = 0.5 cm.
   c) The upper limit is 5.5 cm.
      The lower limit is 4.5 cm.

2. a) The smallest unit of measurement is 0.1 s
   b) The absolute error is half of 0.1 s = 0.05 s
   c) The upper limit is 30.65 s
      The lower limit is 30.55 s.

EXERCISE B

For each of the following measurements give
  a) the smallest unit of measurement,
  b) the absolute error,

  c) the upper and lower limits of the true measurement.
    1. 7 cm   2. 186 m
    3. 17 hours   4. 9.2 kg
    5. 12.8 g   6. 2.683 litres
    7. 1.21 miles   8. 16.5 seconds

Relative error

The absolute error in itself is useful, but it is often more useful to find the **relative error.** This is found by considering the absolute error in relation to the measurement itself.

$$\text{Relative error} = \frac{\text{absolute error}}{\text{measurement}}$$

Example

Find the relative error in giving a length as 7.5 cm

Solution

$$\text{absolute error} = 0.05 \text{ cm}$$

$$\text{relative error} = \frac{0.05}{7.5}$$

$$= \frac{5}{750}$$

$$= \frac{1}{150}$$

This is often expressed as a percentage.

$$\text{percentage error} = \frac{1}{150} \times 100\%$$

$$= \frac{100}{150}\%$$

$$= \frac{2}{3}\%$$

$$= 0.67\%$$

Notice that whereas the absolute error is a quantity of the same kind as the measurement itself, the relative and percentage errors are numbers only.

In calculating absolute, relative and percentage errors, we always calculate the maximum possible errors and so in practice, the errors are likely to be less than those calculated.

EXERCISE C

Calculate for each of the following measurements *a*) the absolute error *b*) the relative error.

1. 150 m
2. 35 kg
3. 1.5 litres
4. 2.3 seconds

Calculate for each of the following measurements *a*) the relative error *b*) the percentage error (to 2 significant figures)

5. 7 cm
6. 3.5 kg
7. 17 days
8. 25.0 g
9. 12.50 m
10. 150.00 litres

The sum and difference of measurements

Example 1
What are the upper and lower limits to the sum of 6.4 cm and 1.2 cm (each given to 2 significant figures).

Solution
6.4 cm lies within the range 6.35 cm and 6.45 cm

and, 1.2 cm lies within the range 1.15 cm and 1.25 cm

Hence, the maximum sum is $6.45 + 1.25$
$$= 7.70 \text{ cm}$$

and, the minimum sum is $6.35 + 1.15$
$$= 7.50 \text{ cm}$$

Note: The apparent sum $(6.4 + 1.2)$ cm = 7.6 cm, has a (maximum) error of 0.10 cm, this being the sum of the absolute errors of the original measurements, each of which was 0.05 cm.

Example 2
Find the limits within which lies the difference of 6.5 kg and 3.25 kg.

Solution
6.5 kg lies within the range 6.45 kg and 6.55 kg

and, 3.25 kg lies within the range 3.245 kg and 3.255 kg

The maximum difference is $6.55 - 3.245$ kg
$$= 3.305 \text{ kg}$$

The minimum difference is $6.45 - 3.255$ kg
$$= 3.195 \text{ kg}$$

Note: The apparent difference $6.5 - 3.25$ kg = 3.25 kg, has a (maximum) absolute error of $(3.305 - 3.25)$ kg or $(3.25 - 3.195)$ kg, both being equal to 0.055 which is the sum of the absolute errors of the original measurements.

EXERCISE D

1. Find the upper and lower limits of the true sum of the measurements given.
   a) 6.5 cm and 8.2 cm
   b) 15 g and 17 g
   c) 17 m and 23 m
   d) 16 litres and 23 litres

2. Without actually calculating the sum write out the absolute errors in adding these measurements:
   a) 320 m and 47 m
   b) 32 kg and 14 kg
   c) 17.5 g and 5.3 g
   d) 4.32 cm and 3.18 cm

3. What are the upper and lower limits of the true differences between the following measurements?
   a) 19 cm and 5 cm
   b) 13 kg and 8 kg
   c) 13.5 m and 6.2 m
   d) 17.6 litres and 14.0 litres

4. Without working out the differences write down the absolute error in subtracting these measurements.
   a) 35 m and 17 m
   b) 230 ml and 155 ml
   c) 17.5 cm and 13 cm
   d) 45 g and 24.5 g

18

The product and quotient of measurements

Example 1
Within what limits does the following product lie? 16 cm × 11 cm

Solution
16 cm lies within the range 15.5 cm and 16.5 cm

and

11 cm lies within the range 10.5 cm and 11.5 cm

$$\text{The maximum product} = 16.5 \times 11.5 \text{ cm}^2$$
$$= 189.75 \text{ cm}^2$$
$$\text{The minimum product} = 15.5 \times 10.5 \text{ cm}^2$$
$$= 162.75 \text{ cm}^2$$

Thus the true product lies between 189.75 cm^2 and 162.75 cm^2

The apparent product is

$$16 \times 11 \text{ cm}^2 = 176 \text{ cm}^2$$

and, $(189.75 - 176) \text{ cm}^2 = 13.75 \text{ cm}^2$

$(176 - 162.75) \text{ cm}^2 = 14.25 \text{ cm}^2$

so that the (maximum) absolute error is
14.25 cm^2

Example 2
Within what limits does this quotient lie?
$$195 \div 13$$

Solution
195 lies within the range 194.5 and 195.5

and, 13 lies within the range 12.5 and 13.5

The maximum quotient is
$$\frac{195.5}{12.5} = 15.64$$
and, the minimum quotient is
$$\frac{194.5}{13.5} = 14.41$$

Thus, the true quotient lies between 15.58 and 14.41

The apparent quotient
$$= \frac{195}{13} = 15$$

so, the absolute error is
$$(15.64 - 15) = 0.64$$
or
$$(15 - 14.41) = 0.59$$

i.e. the maximum absolute error is 0.64

EXERCISE E

Find the upper limits of the following expressions.

1. 32×15 2. $304 \div 16$ 3. $\dfrac{42 \times 13}{17}$

4. 14.5×11.3 5. $201.5 \div 19.3$ 6. $\dfrac{16.3 \times 13.4}{23.7}$

Calculate the absolute error of these expressions.

7. 16×13 8. $210 \div 23$ (figures given correct to 2 significant figures)

9. $436 \div 19$ 10. $\dfrac{420 \times 32}{120}$ (figures given correct to 2 significant figures)

FREQUENCY DISTRIBUTIONS

Variables

When we collect information to be used statistically we want to find out about a particular characteristic of a group of people or objects.

For example, we might want to know the heights of the boys in a class, the number of marks scored in an examination by the pupils of a class or the number of tomatoes on each plant in a certain greenhouse. The particular characteristic in which we are interested, is called the **variable** This variable 'varies' or changes from one member of the group to another.

Variables fall into two distinct types—
discrete and **continuous.**

A **discrete** variable is one which can only have certain definite values, very often whole numbers, e.g. the number of tomatoes on each plant in a greenhouse must be a definite value, in this case a whole number. There might be 2 tomatoes on one plant and 6 on another etc. but there cannot be 2.683 tomatoes on any plant. Some values of discrete variables are not in whole numbers. In shoe sizes there are the 'half' sizes. As a girl's foot grows she may take a size 1, $1\frac{1}{2}$ 2, $2\frac{1}{2}$ but no matter the actual length of the foot, she cannot buy a size 1.75. Shoes can only be bought in certain, distinct sizes.

The various values of a discrete variable can usually be obtained by **counting.** The number of tomatoes per plant can be counted. The number of girls who take a size 2 shoe can be counted.

Let us consider, now, the actual length of a girl's foot as it grows. At age 10 it measured 20 cm and at age 12 it was 25 cm. There is **no length** between 20 cm and 25 cm that the foot did not measure at some time. At one point in time it measured 21.25 cm, at another time it was 23.6847 cm etc. The length of the foot, then, varied continuously —there were no 'gaps' in its length between 20 cm and 25 cm.

A **continuous** variable is one which can take up any value within a certain range. The different values of a continuous variable are usually obtained by some kind of **measurement.**

A boy's height or weight varies continuously and may be found by measuring.

EXERCISE A

State whether the following variables are discrete or continuous.

1. A baby's weight over the first year of its life.

2. The marks of a class in an English examination.

3. The temperature of a person who is ill.

4. The batting scores of a cricket team.

5. The speed of an aeroplane during a journey.

6. The length of a boy's left foot from age 5 to age 10.

7. The shoe sizes of the boy in question 6.

8. The number of goals scored by a football player during one football season.

9. The volume of water used by a town throughout the year.

10. The rainfall in Glasgow over a period of ten years.

11. Air pressures recorded at a weather station.

12. Goals scored by a netball team every week.

Frequency tables (ungrouped data)

The following marks are those scored by 50 pupils in an arithmetic test, the maximum mark being 10.

9	7	5	3	7	8	7	8	6	6
3	5	7	5	8	6	5	6	5	9
3	4	1	6	5	2	3	4	7	8
4	2	7	7	4	6	6	4	10	7
6	5	6	4	5	4	4	5	7	6

Data presented like this as a collection of single facts (a raw score) is not very useful. If you had sat that test and scored the mark of 6, you would like to know how your mark compared with the rest—whether it was a good mark or a poor one. This data must be tabulated. One way of doing this is to write down, in order, the possible marks and count up how many pupils scored each mark. The counting is done by means of tally marks. One stroke is marked down for each time the score occurs, every fifth stroke being drawn diagonally across the previous four, making groups of 5. If the score on the list is crossed out as it is tallied, it helps to avoid confusion. The number of times each score occurs is called the frequency of the score.

Scores in Arithmetic Test (above)

Score	Tally	Frequency
0		0
1	1	1
2	11	2
3	1111	4
4	ⵏ 111	8
5	ⵏ 1111	9
6	ⵏ ⵏ	10
7	ⵏ 1111	9
8	1111	4
9	11	2
10	1	1
		50

You must always check that the total frequency is the same as the number of scores. If it does not come to the same answer, then you must check the tallying.

This table that we have made is called a **Frequency Distribution Table,** since it shows the frequency with which the various marks occur.

An important thing to take note of here is the **range** of the marks—in this case the marks range from 1 to 10. i.e. there is a range of 9 marks.

Now that we have tabulated the data it is much clearer what value a score of 6 has. 6 is, in fact, the score which occurred most often and this 'fashionable' score is called the **mode.**

16 pupils out of 50 or 32% of the group had a score greater than 6, and 24 pupils out of 50 or 48% of the group had a score lower than 6.

It is often important to know how the frequencies of the various scores compare to one another and to the total frequency.

The **relative frequency** of a score is the frequency of this score compared to the total frequency. For example, in the above table, the relative frequency of the score 6 is 10 out of 50, or $\frac{10}{50}$, or 20%, or 0.2.

21

EXERCISE B

1. These are the heights of 40, second year girls, measured to the nearest 2 cm. Make a frequency table and then answer the questions below.

154	144	152	158	158	154	148	150	160	154
156	152	150	154	158	150	154	156	142	154
152	152	154	152	156	148	150	154	154	152
150	158	156	158	158	154	164	148	152	152

a) State whether the variable here is continuous or discrete.
b) What is the range of the heights?
c) What is the modal height?
d) What is the relative frequency of 154 cm?
e) How many girls are 148 cm or less?
f) What percentage of the girls are taller than 158 cm?
g) What percentage of the girls are at least 150 cm tall but not more than 156 cm?

2. List the heights of the pupils in your own class and make a frequency table. Find out what the mode and the range of heights are. Compare your results with those in question 1. Explain any differences or similarities.

3. A pupil collected the following data from a third year class in November 1971.

These are the number of children in the families of the pupils questioned.

1	6	3	2	5	9	6	2	3	6
7	3	2	5	4	3	2	6	5	6
4	8	10	5	6	1	10	6	3	

Make a frequency table.

a) Is the variable continuous or discrete?
b) What is the mode?
c) What kind of family is not taken into account in this table? Why not?
d) How many of the pupils in this class belong to families with more than 6 children?
e) What percentage of the pupils are in families with 3 or less children? (answer correct to 1st decimal place).

4. Make a list of the number of children in the families of the pupils in your own class and make a frequency table.

Compare your results with those in the last question. Point out any similarities or differences.

5. These are the results of the Scottish Football matches (First and Second Division) one Saturday in the 1968–69 season.

2—0	2—1
1—1	0—2
2—4	2—0
3—1	1—3
2—1	1—3
3—1	0—2
1—0	2—1
4—0	5—1
2—0	2—4

a) Form a frequency table of the number of goals scored by each team.
b) Is this a continuous or discrete variable?
c) What is the range of goals scored?
d) What is the mode?
e) What is the relative frequency of the mode?
f) What percentage of the teams scored 3 or more goals?

6. Repeat the last questions with your own data about recent football results.

Histograms (ungrouped data)

One of the best ways of representing a frequency distribution graphically is by means of a **histogram**.

This is the frequency table that we made of the marks in the arithmetic test earlier in the chapter.

Score	0	1	2	3	4	5	6	7	8	9	10
Frequency	0	1	2	4	8	9	10	9	4	2	1

Fig. 3.1 Arithmetic Test

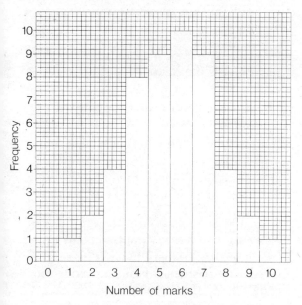

Number of marks

As you see, a histogram is a column graph in which the areas of the rectangles are proportional to the frequencies.

The columns can be drawn without leaving any spaces between because there is a regular scale along the horizontal axis.

The values of the variable are always shown on the horizontal axis and the frequencies on the vertical axis.

EXERCISE C

1. The histogram below (Fig. 3.2) shows the number of points gained by the teams of the Second Division of the English football league at one stage of the 1968–69 season.

a) How many clubs are in this league?

b) What is the range of the points?

c) What is the mode number of points?

d) What is the relative frequency of 16 points?

e) How many clubs gained 22 or more points?

f) How many clubs gained less than 15 points?

Fig. 3.2 Points Scored By Teams of English Second Division

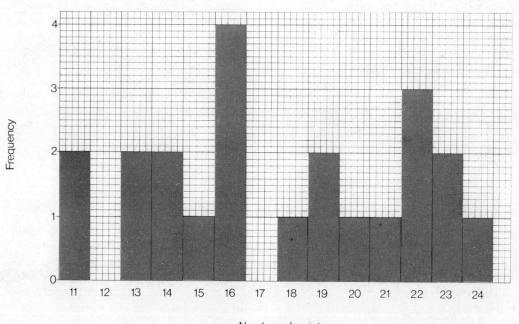

Number of points

2. The following histogram illustrates the
distribution of the shoe sizes of all boys
aged 12–13 years in a school one January.

Fig. 3.3 Shoe Sizes of Boys Aged 12–13 Years

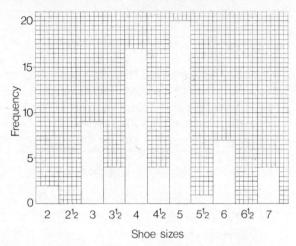

Shoe sizes

a) Do you notice anything unexpected
   about the distribution of the
   frequencies? Can you suggest any
   reason for this?
b) What is the total frequency of the
   distribution?
c) What is the modal shoe size?
d) What is the relative frequency of the
   modal size?
e) What is the relative frequency of size 6?

3. Draw histograms for the frequency
distributions you tabulated for Exercise B.

4. The numbers of words in each line of a
page of the book *Red Gauntlet* by Sir Walter
Scott are shown below. Form a frequency
table and draw a histogram of the
distribution.

11	10	6	8	12	1	9	5
7	8	8	8	8	9	10	8
8	8	3	7	13	5	10	9
7	9	6	8	6	8	11	
9	9	8	9	11	3	6	

a) How many lines are in the page?
b) What is the modal number of words
   per line?
c) What is the range of the distribution?
d) What is the relative frequency of 9
   words?
e) How many lines have more than 9
   words?
f) How many lines have less than 5
   words?
g) How many lines have no more than 8
   words and no less than 6?

5. Choose any page at random from a novel
and count the number of words in each line.
Construct a frequency table and draw the
histogram for the distribution.

6. A class of first year pupils was given a
speed and accuracy test consisting of ten
sums—addition, subtraction, multiplication
and division of number. The results obtained
were as follows:

No. of sums correct	0	1	2	3	4	5	6	7	8	9	10
Frequency	0	1	2	6	11	5	2	2	1	2	0

Draw a histogram of the distribution.
a) How many pupils took the test?
b) What was the modal number of correct
   sums?
c) What was the range of the marks?
d) What percentage of the pupils had more
   than 5 sums correct?

Frequency polygons (ungrouped data)

Another way of representing a frequency
distribution graphically is by means of a line
graph called a **frequency polygon**. A
frequency polygon is drawn by joining, with
straight lines, the mid-points of the tops of
the columns of the histogram. Figure 3.4a
shows a frequency polygon superimposed on

the histogram. Figure 3.4(b) shows the polygon by itself. To draw the polygon without drawing the histogram first, the points are plotted where the mid points of the columns would have been.

Fig. 3.4 (a) Arithmetic Test

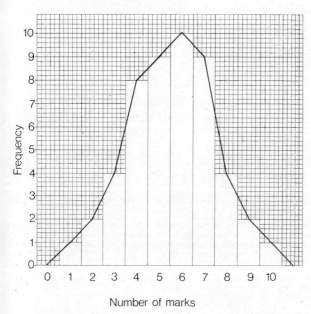

Number of marks

Fig. 3.4 (b) Arithmetic Test

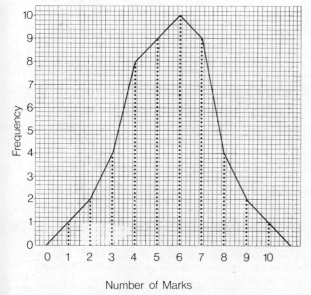

Number of Marks

Since the area under the polygon must be equal to the area contained by the columns of the histogram, the polygon ought to be finished off by continuing the line at each end to the horizontal axis to where the next score would have been found.

Frequency polygons are very useful if we want to compare two distributions. It is difficult to draw two histograms on the same graph so that the distributions are clearly seen. However, it is very easy to draw two or more polygons on the same graph. The only drawback to this kind of comparison is that the total frequencies of the distributions must be the same, in order to give a fair comparison.

EXERCISE D

Fig. 3.5 Goals Scored By Teams of The English Football League One Saturday

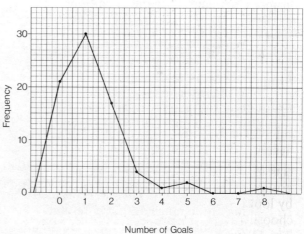

Number of Goals

1. This polygon shows the number of goals scored by the teams of the four divisions of the English Football League one Saturday in the 1968–69 season.
   a) How many teams scored 1 goal?
   b) How many teams scored more than 2 goals?
   c) How many teams played that Saturday?

25

2. Draw frequency polygons on the histograms you drew for Exercise C.

3. Draw polygons by themselves for the distributions in Exercise C.

4. Fifty pupils were given two tests in successive weeks. The results were as follows:

Marks		0	1	2	3	4	5	6	7	8	9	10
Frequency	1st test	1	4	7	15	12	5	4	2	0	0	0
	2nd test	0	0	0	3	4	7	20	8	5	2	1

On the same graph draw frequency polygons for each distribution. (Use a solid line for the first test and a broken line for the second)

   a) What was the modal mark in each test?

   b) Comment on the way the marks are distributed. What reason might there be for it?

5. Measure the heights of 40 boys and 40 girls in the same year (to the nearest 2 cm). Form frequency tables and draw frequency polygons for each distribution on the same graph. Comment on the distributions. ALTERNATIVELY—measure the heights of 40 boys (or girls) in one year and 40 boys (or girls) in another year and draw polygons for each distribution on the same graph.

Frequency tables (grouped data)

Sometimes the data we are considering has such a large range of scores that it is necessary to collect the scores into groups or classes. The **class interval** is the size of the group chosen. The class interval is decided by looking at the range of the scores and choosing the interval so that there are about 8 to 14 groups.

80	93	63	74	51	60	61	53	69	54
50	43	32	70	40	30	55	57	59	55
4	50	42	47	53	52	67	44	35	48
37	56	39	52	12	44	43	58	49	32
65	64	62	73	21	21	86	67	75	68

The figures at the bottom of the last column are a list of the marks of 50 boys in an examination. First we look through the list and pick out the highest and lowest scores. These are 93 and 4. A good way of grouping these marks is in tens—starting with the group 0–9, then 10–19, 20–29 etc. This will give us 10 groups. The scores are tallied in exactly the same way as for ungrouped data.

Mark	Tally	Frequency
0– 9	1	1
10–19	1	1
20–29	11	2
30–39	1111 1	6
40–49	1111 1111	9
50–59	1111 1111 1111	14
60–69	1111 1111	10
70–79	1111	4
80–89	11	2
90–99	1	1

With grouped data we cannot pick out a single mark as the mode, but we can pick out the **modal class** or **group.** In this frequency distribution, the modal class is the group (50–59) marks.

One point that should be noted here is that a certain amount of accuracy is lost when data is grouped. For instance, from the table, we know that there are 6 scores between 30 and 39. Unless we have the original raw scores we do not know that the actual scores are 30, 32, 32, 35, 37 and 39.

Class boundaries

When selecting a grouping care must be taken to ensure that no gaps are left between the classes and also that the classes are not allowed to overlap. In the above example it would be useless to take classes 0–10, 10–20, 20–30, etc. as here we would not know whether to place a score of 10 in the first or second class. Neither of these difficulties will occur if we have a clear understanding of the nature of our data and of our class boundaries. In the above example the actual marks in the second class are 10–19, but we say that the boundaries enclosing the class are 9.5 and 19.5. If the first few classes are written in a row it helps to make this clearer.

Class	0–9	10–19	20–29	30–39	
Boundaries		9.5	19.5	29.5	39.5

Thus 9.5 is the boundary separating the first and second classes i.e. 9.5 is the upper boundary of the first class and the lower boundary of the second class. We are in fact taking as our class boundary a measure which is half way between the largest measure we wish to include in the one class and the smallest we wish to include in the next. With most types of data this is quite straightforward but occasionally, especially when dealing with age, care must be taken e.g. the group of 13 year old children have ages ranging from 13 years exactly, up to but not including 14 years.

The mid-point of the class is now taken as the average of the upper and lower boundaries of the class. In the above example dealing with marks the mid-point of the second class is 14.5 and for the class of 13 year old children the mid-point of the class is 13.5 years. The mid-point of the class interval is important, as you will see later,

because it is often used to stand for the whole group.

EXERCISE E

1. Form a frequency table of the test marks of 40 pupils shown below, using class intervals of 5 marks starting at 1–5, 6–10 etc.

20	19	18	6	25	15	16	30	17	25
2	22	5	22	23	8	15	7	7	12
11	26	21	12	14	28	9	18	13	19
16	31	38	20	33	10	42	16	47	13

a) What is the mid-point of the 5th class?
b) What is the upper boundary of the 5th class?
c) What is the lower boundary of the 2nd class?
d) What is the modal class?
e) What is the relative frequency of the third class?
f) What percentage of the pupils scored more than 25 marks?
g) What percentage of the pupils scored more than 10 marks but less than 26?

2. Construct a frequency table for the examination marks of your own class in any subject and group the marks in suitable class intervals.

3. The table at the foot of the page shows the infant mortality rates (per 1000 live births) in Scotland for each quarter year, over several years. Construct a frequency table of the quarterly rates choosing suitable class intervals.

Year	1954	–55	–56	–57	–58	–59	–60	–61	–62	–63	–64
1st Quarter	38	36	31	32	32	33	29	30	30	29	27
2nd ,,	30	29	26	27	27	26	26	25	26	24	23
3rd ,,	24	26	27	27	25	25	24	23	24	21	22
4th ,,	31	31	29	29	27	29	27	26	26	28	23

a) What is the mid-point of the 1st class?
b) What is the upper boundary of the 3rd class?
c) What is the lower boundary of the 5th class?
d) What is the modal class?
e) What is the relative frequency of the 5th class?

4. The following figures were the number of rain days (days in which rain fell) recorded at selected stations all over the U.K. in one year:

260	234	209	241	243	268	185
241	253	228	199	158	236	209
173	187	191	207	199	199	172
178	157	188	148	183	160	152
197	171	265	226	235	226	233
180	182	211	202	185	206	209
182	188	164	156	160	250	220
187	195	215	177	191	225	164
209	206	194	205	195	198	177
175	180					

Construct a frequency table using class intervals of 10 starting at 140–149.

a) What is the mid-point of the 4th class?
b) What is the lower boundary of the 6th class?
c) What is the upper boundary of class 10?
d) What is the modal class?
e) What number of stations recorded less than 200 rain days?
f) What percentage of stations recorded at least 170 rain days and not more than 209?

Histograms (grouped data)

To represent graphically frequency distributions with grouped data, we again use histograms. However with grouped data, we must be very careful how we use the horizontal scale.
Below is the frequency distribution of marks that we tallied in the previous section.

Mark	0–9	10–19	20–29	30–39	40–49	50–59	60–69	70–79	80–89	90–99
Frequency	1	1	2	6	9	14	10	4	2	1

Fig. 3.6 Examination Marks

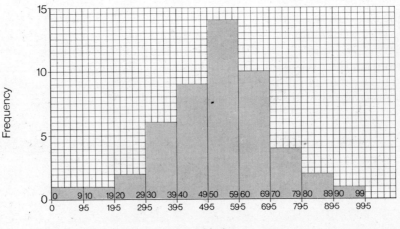

Take particular note of the scale used on the horizontal axis in Figure 3.6. On the actual scale, the class boundaries are used, and the groups are shown above. As you become used to drawing these histograms you should start to leave out the groups and simply show the boundaries.

EXERCISE F

1. The histogram shows the distribution of the rainfall (in cm) recorded at selected stations all over Britain, one year.

Fig. 3.7 Rainfall Recorded in Britain For a Year

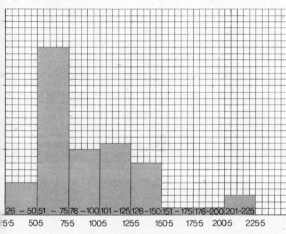

Rainfall in cm.

a) How many stations were there?
b) What is the upper boundary of the 1st class?
c) What is the lower boundary of the 5th class?
d) What is the mid-point of the 3rd class?
e) What is the modal class?
f) What is the relative frequency of the modal class?
g) What percentage of the stations recorded a rainfall of more than 125 cm?
h) What percentage of the stations recorded a rainfall of 50 cm or less?

2. The distribution of the weights (in kg) of a number of boys is shown in the histogram Figure 3.8.
   a) How many boys were weighed?

Fig. 3.8 Weights of Boys (in Kg)

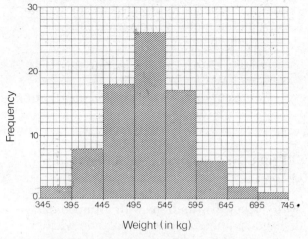

Weight (in kg)

b) What is the upper boundary for the 3rd class?
c) What is the lower boundary for the 3rd class?
d) What weights are in the 3rd class?
e) What weights are in the 5th class?
f) What is the mid-point of the 4th class?
g) What is the modal class?
h) What is the relative frequency of the modal class?

3. Draw histograms for all the distributions you tabulated in Exercise E.

4.

Age	Frequency
0– 4	37
5– 9	47
10–14	19
15–19	9
20–24	5
25–29	4
30–34	11
35–39	13
40–44	8
45–49	11
50–54	20
55–59	23
60–64	33
65–69	30
70–74	30
75–79	28
80–84	28
85+	10

The table on the previous page shows the number of pedestrians killed in one year in Scotland in motor vehicle accidents, according to age in years.

Draw a histogram to illustrate this, without any further grouping.

Comment on the shape of the distribution and give reasons for it.

Frequency polygons (grouped data)

Frequency polygons are used to show frequency distributions with grouped data, as line graphs. To draw a polygon we join up with straight lines the mid-points of the tops of the columns of the histogram.

Example 1

The marks of some pupils in a test are shown below grouped in class intervals of 5.

You can simply draw this like any line graph, plotting the mid-points against the corresponding frequency—see Figure 3.9 (b).

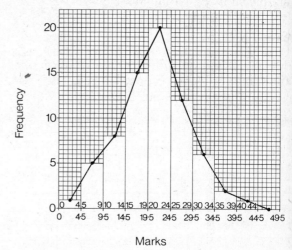

Fig. 3.9 (a) Test Results

Mark	0–4	5–9	10–14	15–19	20–24	25–29	30–34	35–39	40–44	45–49
Frequency	1	5	8	15	20	12	6	2	1	0

Figure 3.9 (a) shows the polygon of this distribution drawn on top of the histogram.

Figure 3.9 (b) shows the polygon by itself.

If you study Figure 3.9 (a) you will see that the key points for the polygon are actually above the mid-points of the class intervals, and to draw the polygon on a graph by itself you need only plot the mid-points of the intervals against the corresponding frequencies.

The mid-points of the intervals are shown below.

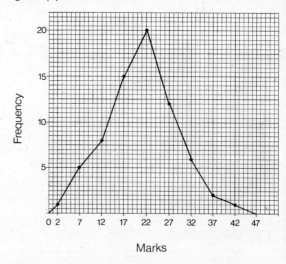

Fig. 3.9 (b) Test Results

Mid-point of interval	2	7	12	17	22	27	32	37	42	47
Frequency	1	5	8	15	20	12	6	2	1	0

Example 2
Draw the frequency polygon for the following
distribution of examination marks.

Mark	0–9	10–19	20–29	30–39	40–49	50–59	60–69	70–79	80–89	90–99
Frequency	1	1	2	6	9	14	10	4	2	1

Solution
First find the mid-points of the class intervals

Mid-point	4.5	14.5	24.5	34.5	44.5	54.5	64.5	74.5	84.5	94.5
Frequency	1	1	2	6	9	14	10	4	2	1

Fig. 3.10 Examination Marks

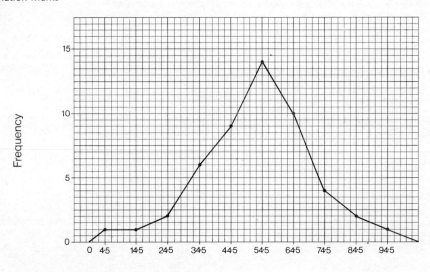

EXERCISE G

1. Draw frequency polygons superimposed
on the histograms you drew for Exercise F.

2. Draw polygons by themselves (adjusting
your scale accordingly) for the distributions
in Exercise F.

3. The heights of 50 second year boys were
measured to the nearest cm in October 1967.
In October 1968 the heights of 50 boys in
the second year were again measured (to the
nearest cm). These heights are shown below.
   Draw, on the same graph, frequency
polygons for each distribution.
   Comment on any differences you find.

Height (cm)		116–120	121–125	126–130	131–135	136–140	141–145	146–150	151–155	156–160	161–165	166–170	171–175
Freq.	1967	2	0	0	0	0	5	7	14	11	8	2	1
	1968	0	0	0	0	1	3	10	20	12	3	1	0

4. The number of words in each of the first hundred sentences of two books were counted. One book was 'Moby Dick' by Herman Melville (the story of the giant whale) first published in 1851. The other book was a modern, run of the mill, detective story first published some hundred years later.

The distributions for the number of words are shown below.

Draw polygons for the distributions on the same graph and comment on the way the distributions are dispersed.

What does the graph indicate to you about the 'readability' of the two books.

Moby Dick

No. of words	Frequency
1– 10	30
11– 20	23
21– 30	22
31– 40	15
41– 50	4
51– 60	0
61– 70	3
71– 80	1
81– 90	1
91–100	0
101–110	0
111–120	1

Detective Story

No. of words	Frequency
1–10	62
11–20	24
21–30	10
31–40	2
41–50	0
51–60	1
61–70	1

5. Choose 2 novels (*not* textbooks) and count the number of words in each of the first 100 sentences and repeat question 6.

6. Shown below are the number of hours of bright sunshine recorded in 52 selected stations all over Scotland one year, in the months of December and June.

Tabulate the data for each month grouping in suitable class intervals, and draw a frequency polygon for each month on the same graph.

Compare and contrast the two distributions.

December — No. of hours of bright sunshine

16	25	41	20	35	20	16	8	38	23
25	38	38	41	34	39	24	47	45	17
42	44	47	45	51	35	37	51	47	42
42	31	39	55	45	42	33	27	35	40
28	35	39	36	14	14	40	44	43	34
50	43								

June — No. of hours of bright sunshine

163	153	164	137	177	160	158	148
171	152	169	180	179	152	164	157
181	198	188	163	195	195	209	212
195	204	197	215	206	192	201	149
186	234	192	177	144	120	186	173
176	163	177	175	163	148	161	172
150	136	150	151				

Cumulative frequency

It is often useful to know how many boys have scored more than a certain mark in an examination or how many girls are less than a particular height. An easy way to answer this type of question is by making a **cumulative frequency** table. A frequency table is converted to a cumulative frequency table by adding each frequency to the total of its predecessors.

Example 1
The marks of 30 pupils in a test are shown below (a).

The cumulative frequency table for these marks is shown in (b).

(a)

Mark	Frequency
0	0
1	2
2	3
3	4
4	6
5	8
6	5
7	1
8	1
9	0
10	0

(b)

Mark		Cumulative Frequency
0		0
up to and including 1	(0+2)	2
,, ,, ,, ,, 2	(2+3)	5
,, ,, ,, ,, 3	(5+4)	9
,, ,, ,, ,, 4	(9+6)	15
,, ,, ,, ,, 5	(15+8)	23
,, ,, ,, ,, 6	(23+5)	28
,, ,, ,, ,, 7	(28+1)	29
,, ,, ,, ,, 8	(29+1)	30
,, ,, ,, ,, 9	(30+0)	30
,, ,, ,, ,, 10	(30+0)	30

From the cumulative frequency table we can see, for example, that

23 pupils scored 5 or less marks.
9 pupils scored less than 4 marks.

Example 2
In the next column (a) is the frequency table we constructed a few pages ago of the marks of 50 boys in an examination.

(b) shows the cumulative frequency table for the distribution.

(a)

Mark	Frequency
0– 9	1
10–19	1
20–29	2
30–39	6
40–49	9
50–59	14
60–69	10
70–79	4
80–89	2
90–99	1
	50

(b)

Mark		Cumulative Frequency
up to and including 9		1
,, ,, ,, ,, 19	(1+1)	2
,, ,, ,, ,, 29	(2+2)	4
,, ,, ,, ,, 39	(4+6)	10
,, ,, ,, ,, 49	(10+9)	19
,, ,, ,, ,, 59	(19+14)	33
,, ,, ,, ,, 69	(33+10)	43
,, ,, ,, ,, 79	(43+4)	47
,, ,, ,, ,, 89	(47+2)	49
,, ,, ,, ,, 99	(49+1)	50

Instead of writing
'up to and including 9'
,, ,, ,, ,, 19', etc.,
we can show the mark for the cumulative frequency as

0– 9
0–19
0–29 etc.

From the above cumulative frequency we can answer these questions readily.

1. How many boys scored less than 50 marks? Answer—19

2. How many boys scored more than 59 marks? Answer—17 (50–33)

Height (cm)	146	148	150	152	154	156	158	160	162	164	166	168
Frequency	1	0	1	3	3	4	3	5	6	1	2	1

EXERCISE H

1. At the top of the page are the heights of a class of 30 fourth year girls measured to the nearest 2 cm.
Construct a cumulative frequency table and answer the following questions
   a) How many of the girls are less than 150 cm tall?
   b) What percentage of the girls are at least 158 cm in height?
   c) How many girls are over 164 cm?

2. The frequency distribution for the times of some telephone calls was as follows.

Time (s)	Frequency
0– 20	0
21– 40	2
41– 60	4
61– 80	7
81–100	11
101–120	28
121–140	20
141–160	16
161–180	10
181–200	6
201–220	3

   a) Construct a cumulative frequency table for the distribution.
   b) What number of calls lasted a minute or less?
   c) What percentage of the calls lasted more than 2 minutes?

3. The frequency distribution below shows the weights of fourth year boys measured to the nearest kg.

Construct a cumulative frequency table.
   a) What was the total number of boys weighed?
   b) How many boys weighed less than 50 kg?
   c) How many boys weighed 60 kg or more?

4. In a major Golf Tournament, one September, the leading ten players handed in the following scores (4 rounds each).

70	69	68	69	66	68	71	71	71	69
70	69	71	74	75	68	70	74	72	68
69	70	72	68	70	73	69	67	75	73
66	68	67	68	68	70	70	69	65	71

   a) Construct a frequency table for the scores.
   b) Convert this to a cumulative frequency table.
   c) How many rounds of less than 70 were there?
   d) How many rounds of more than 71 were there?

Cumulative frequency curves

When we draw the graph of a cumulative frequency distribution we obtain a curve which has a characteristic shape. This curve is called a **cumulative frequency curve** (or **ogive**—from a term used in architecture for this shape of curve).
   Figure 3.11 shows the cumulative frequency curve for example 1 of the last section.

Weight (kg)	35–39	40–44	45–49	50–54	55–59	60–64	65–69	70–74	75–79	80–84
Frequency	1	2	3	8	8	4	1	0	1	1

Fig. 3.11 Test Results

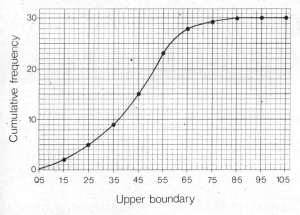

Fig. 3.12 Examination Results

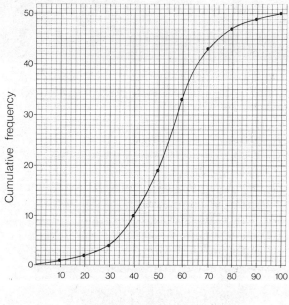

Marks—Upper boundary

Mark	Cumulative Frequency
0	0
0– 1	2
0– 2	5
0– 3	9
0– 4	15
0– 5	23
0– 6	28
0– 7	29
0– 8	30
0– 9	30
0–10	30

To plot the points of the graph, the cumulative frequency is drawn against the upper boundary of the class
e.g.
2 is plotted against the mark 1.5
5 against 2.5
9 ,, 3.5
15 ,, 4.5

Figure 3.11 shows the cumulative frequency curve of Example 2 of the last section.

Mark	Cumulative Frequency
0– 9	1
0–19	2
0–29	4
0–39	10
0–49	19
0–59	33
0–69	43
0–79	47
0–89	49
0–99	50

To plot the points of this graph we plot the cumulative frequency against the upper boundary of the class.
e.g.
1 is plotted against the mark 9.5
2 against 19.5
4 ,, 29.5
10 ,, 39.5

Note: The horizontal scale is marked off in tens but the points are plotted at 9.5, 19.5 etc.

As with any line graph, 2 or more curves may be drawn on the same graph, for purposes of comparison.

EXERCISE I

Draw cumulative frequency curves for all cumulative frequency distributions you tabulated in Exercise H.

35

4 MEASURES OF CENTRAL TENDENCY

Data can be more easily understood when it is tabulated in an orderly fashion in a frequency distribution and then shown graphically. But one single value which characterises the group and which is most easily grasped by the mind, is the **average** of the group.

There are three 'averages' used in statistics:

1. The Arithmetic Mean
2. The Median
3. The Mode

The arithmetic mean

This is the 'average' used in arithmetic but in Statistics we must be careful always to call it the mean.

To find the mean of set of scores, we simply add up the scores and divide by the number of scores.

Example
Find the mean of the following numbers.
1, 3, 5, 7, 9, 11, 13, 15.

Solution

$$\text{The mean} = \frac{\text{Sum of the scores}}{\text{number of scores}}$$

$$= \frac{1+3+5+7+9+11+13+15}{8}$$

$$= \frac{64}{8}$$

$$= 8$$

The formula we use to calculate the mean is,

$$\bar{X} = \frac{\sum X}{N}$$

where $\bar{X}$ = mean

$\sum$ means sum of

X represents the single scores

N represents the number of scores.

The median

When a number of scores are arranged in **numerical order,** the median score is the 'middle' score having the same number of scores above it as below. When there is an odd number of scores this is easy as there is a 'middle' score but when there is an even number of scores there is no single 'middle' score and we define the median as the average of the two 'middle' scores.

Example 1
Find the median of the following golf scores.
72, 68, 65, 70, 75, 79, 73

Solution
First arrange the scores in numerical order.

65, 68, 70, |72,| 73, 75, 79,
median

72 is the median since it has 3 scores above it and 3 below it.

Example 2

Find the median of these boys' heights (measured to the nearest cm)

148, 161, 167, 157, 162, 154.

Solution

First arrange in numerical order.

148, 154, 157, | 161, 162, 167.

median

This time there is an even number of scores and the median is the average of 157 and 161, i.e. the median is

$$\frac{157 + 161}{2} = 159.$$

The mode

The mode, as we saw in the last chapter, is the score which occurs most frequently.

EXERCISE A

1. Find the mean, median and mode of the following sets of data.

   a) 15, 11, 6, 3, 2, 14, 13, 7, 10, 11, 9.

   b) 24, 21, 20, 25, 21, 27.

   c) 5, 2, 2, 1, 3, 3, 3, 4, 4, 3, 3, 4, 2, 3, 2, 4.

2. In successive rounds a golfer took the following numbers of strokes

      90, 69, 70, 70, 73, 71, 80.

Which of the three averages—mean, median or mode—would he prefer to call his 'average' score?

3. The heights (to the nearest cm) of some boys are given below.

162, 159, 162, 167, 165, 158, 166, 164, 158, 160.

What is a) the median height,

     b) the modal height,

     c) the mean height?

4. Make a list of 5 numbers, 4 of which are smaller than the mean.

5. Make a list of 6 numbers, 5 of which are greater than the mean.

6. If you were handed a box of used pencils of varying lengths, what would be the easiest average—mean, median or mode—to find, and what would be the hardest?

7. In his savings account a boy had the following balances at the ends of 6 successive months. What was his mean balance and his median balance?

   £6.50, £9.75, £4.75, £5.90, £8.40, £5.50.

Averages from frequency distributions

In the previous exercise, the mean, median and mode were being obtained from raw scores, but we often want to know these averages from data which can be tabulated into, or is already in the form of, a frequency distribution.

The mean from a frequency distribution (ungrouped data)

Example

Calculate the mean of the following test marks of 50 pupils (possible mark being 10)

   a) without tabulating them,

   b) by constructing a frequency table.

1,	5,	5,	3,	4,	6,	5,	6,	4,	7
7,	6,	6,	7,	4,	7,	2,	3,	5,	6
8,	4,	2,	6,	7,	4,	7,	5,	6,	4
7,	9,	7,	3,	5,	8,	3,	5,	4,	8
4,	6,	6,	10,	7,	6,	5,	9,	8,	5

Solution

a)   The mean $\overline{X} = \dfrac{\sum X}{N}$

$$= \frac{277}{50}$$

$$= 5.54$$

37

b)

Score	Tally	Frequency
0		0
1	1	1
2	11	2
3	1111	4
4	1111 111	8
5	1111 1111	9
6	1111 1111	10
7	1111 1111	9
8	1111	4
9	11	2
10	1	1
		$\sum f = 50$

$$\overline{X} = (0 \times 0) + (1 \times 1) + (2 \times 2) + (4 \times 3)$$
$$+ (8 \times 4) + (9 \times 5) + (10 \times 6) + (9 \times 7)$$
$$\frac{+ (4 \times 8) + (2 \times 9) + (1 \times 10)}{50}$$

$$= \frac{277}{50}$$

$$= 5.54$$

To obtain 277 what we are doing is multiplying the frequency of each score by the score and adding the results.

An easier way of doing this is to add another column to the frequency table in which we multiply the frequency (f) by the score (X)

Score (X)	Frequency (f)	$f \times X$	
0	0	(0 × 0)	0
1	1	(1 × 1)	1
2	2	(2 × 2)	4
3	4	(4 × 3)	12
4	8	(8 × 4)	32
5	9	(9 × 5)	45
6	10	(10 × 6)	60
7	9	(9 × 7)	63
8	4	(4 × 8)	32
9	2	(2 × 9)	18
10	1	(1 × 10)	10
	$\sum f = 50$	$\sum fX = 277$	

$$\overline{X} = \frac{\sum fX}{\sum f} \quad (\sum fX = \text{sum of frequency} \times \text{score}$$
$$\sum f = \text{sum of frequencies})$$

$$= \frac{277}{50}$$

$$= 5.54$$

EXERCISE B

1. This frequency table shows the number of goals scored by the teams of the First and Second Division of the English Football League on Saturday 5th October 1968.

Number of goals	0	1	2	3	4	5
Frequency	15	16	4	6	2	1

Calculate the mean number of goals scored.
2. Construct a frequency table from the goals scored by the English First and Second Division teams last Saturday and calculate the mean number of goals.
3. The following table shows the number of children per family in the families of the pupils in a first year class.

Number of children	1	2	3	4	5	6	7	8	9
Frequency		2	4	10	6	6	3	2	1	0

Calculate the mean number of children per family.
4. Construct a frequency table of the number of children in the families of the pupils in your class and calculate the mean. (If you have already constructed such a table for question 4 in Exercise B of Chapter 3, then use it again.)
5. The following numbers are the number of words in each line of a page chosen from the novel *Westward Ho!* by Charles Kingsley.

7	7	7	10	13	8	10	6	9	12
11	1	12	13	2	11	10	10	13	9
12	1	10	8	12	12	12	12	3	7
10	12	9	10	12	11	12	9	2	11
8	13								

Construct a frequency table and then calculate the mean number of words per line.

6. Repeat the last question by choosing any novel and counting the number of words in each line of a page chosen at random.

The mean from a frequency distribution (grouped data)

Example

Calculate the mean of the following test marks (possible mark 50)
a) without grouping
b) by grouping in class intervals of 5.

4	23	35	27	32	15	29	19	25	18
36	11	33	12	6	30	20	24	27	25
16	40	21	35	20	17	9	30	14	29
31	26	23	25	21	32	23	18	26	11
28	43	24	43	24	10	37	22	34	23

Solution

a)

$$\bar{X} = \frac{\sum X}{N}$$

$$= \frac{1206}{50}$$

$$= 24.12$$

b)

Mark	Tally	Frequency
1–5	1	1
6–10	111	3
11–15	₩₩t	5
16–20	₩₩t 11	7
21–25	₩₩t ₩₩t 111	13
26–30	₩₩t 1111	9
31–35	₩₩t 11	7
36–40	111	3
41–45	11	2
46–50		0
		$\sum f = 50$

When we have a frequency distribution with grouped data, we use the mid-point of the interval to stand for the group, and as before, we add a frequency × score column to the table.

To calculate the mean, we again use the formula

$$\bar{X} = \frac{\sum fX}{\sum f}$$ (This time X is the mid-point of the interval.)

Mark	Mid-point of interval (X)	Freq. (f)	fX
1–5	3	1	3
6–10	8	3	24
11–15	13	5	65
16–20	18	7	126
21–25	23	13	299
26–30	28	9	252
31–35	33	7	231
36–40	38	3	114
41–45	43	2	86
46–50	48	0	0
		$\sum f = 50$	$\sum fX = 1200$

$$\bar{X} = \frac{\sum fX}{\sum f}$$

$$= \frac{1200}{50}$$

$$= 24.0$$

Note: When the data is grouped the mean is slightly different from that obtained when we use the raw scores. This is because of the slight loss of accuracy any time data is grouped. The more correct mean is that obtained from the raw scores.

The mean by using a calculating machine

The calculation of a mean from a frequency table can be carried out quickly by using a machine. Set 1 at the left hand end of the setting register and in turn the various scores or mid-points of the intervals at the right hand end of the setting register. Multiply each score by its appropriate frequency, clearing the counting register after each multiplication but allowing the results to accumulate in the product register. You will finish with the sum of the frequencies at the left hand end and the sum of the products at the right hand end of the product register.

EXERCISE C

1. These are the number of deaths due to railway accidents in Scotland in 1964, according to the age of the deceased person (in years):

Age	5–9	10–14	15–19	20–24	25–29	30–34	35–39
Frequency	1	1	3	2	1	1	1

Age	40–44	45–49	50–54	55–59	60–64	65–70
Frequency	1	1	1	5	2	1

Calculate the mean age of death.

2. This table shows the marks gained by 50 pupils in an arithmetic examination. Calculate the mean mark.

Mark	20–29	30–39	40–49	50–59	60–69	70–79	80–89
Frequency	1	4	8	15	16	5	1

3. The age distribution of workers in a factory was as follows. Calculate the mean age.

Age (years)	Frequency
16–20	2
21–25	10
26–30	12
31–35	17
36–40	15
41–45	14
46–50	9
51–55	8
56–60	8
61–65	5

4. These figures show the number of hours of bright sunshine recorded in 52 places in Britain one year for the month of October.

50	81	75	90	62	80	92	75	77	74
86	99	75	63	67	73	95	75	71	74
93	94	76	68	79	81	105	83	81	75
71	82	79	76	69	85	75	80	81	47
99	68	71	74	74	85	65	71	74	80
69	68								

Construct a frequency table using appropriate class intervals and calculate the mean number of hours of bright sunshine.

An alternative method of calculating the mean using an assumed mean

This method of calculating the mean cuts down considerably on the amount of multiplication but care must be taken in using it.

In this method, we choose an estimated or assumed mean and work from it as shown below. We find the deviation of each score from the assumed mean and then find the mean of these deviations and apply this to the assumed mean to obtain the true mean.

Example 1
Here are the examination marks of 8 pupils:
56, 58, 59, 60, 62, 63, 64, 65.
Find the mean mark, using an assumed mean.

Solution

Mark	Assumed Mean	Deviation from Assumed Mean (d)
56		−4
58		−2
59		−1
		−7
60	60	0
62		+2
63		+3
64		+4
65		+5
		+14
		$\sum d = -7 + 14$
		$= +7$

$$\text{Mean} = \text{Assumed Mean} + \frac{\sum d}{N}$$

$$= 60 + \frac{7}{8}$$

$$= 60 + 0.875$$

$$= 60.875$$

The approximate mean mark = 60.9

Example 2

The table below shows the heights of 26 pupils measured to the nearest centimetre. Find the mean height using an assumed mean of 152 cm.

Solution

(When there is a frequency for the scores, then a frequency × deviation column must be added to the table. This time we find the mean of the f × d column.)

$$\text{Mean} = \text{assumed mean} + \frac{\sum fd}{\sum f}$$

$$= 152 + \frac{9}{26}$$

$$= 152 + 0.35$$

$$= 152.35$$

The approximate mean height = 152.4 cm.

Grouped data

With grouped data we use the mid-point of the interval to stand for the whole group and we can further simplify the calculation by using as the unit of deviation from the assumed mean the band width. Here then d will measure band widths and when calculating the mean we must multiply $\sum fd$ by the band width to get the total deviate.

Height (cm)	Frequency (f)	Assumed Mean	Deviation from Assumed Mean (d)	f × d
148	1		−4	−4
149	1		−3	−3
150	2		−2	−4
151	3		−1	−3
				−14
152	6	152	0	0
153	7		+1	+7
154	3		+2	+6
155	2		+3	+6
156	1		+4	+4
	$\sum f = 26$			+23
				$\sum fd = -14 + 23$
				$= +9$

Example 3

The rateable values of eighty houses in one street were found and the results shown below. Calculate the mean rateable value using an assumed mean.

Rateable value (£)	50–59	60–69	70–79	80–89	90–99	100–109	110–119
Frequency	2	10	16	24	16	11	1

Solution

Rateable Value (£)	Frequency (f)	Assumed Mean	d	fd
50– 59	2		−3	−6
60 69	10		−2	−20
70– 79	16		−1	−16
				−42
80– 89	24	84.5	0	0
90– 99	16		+1	+16
100–109	11		+2	+22
110–119	1		+3	+ 3
				+41
	Σ f = 80			Σ fd = −42+41 = −1

$$\text{Mean} = \text{Assumed mean} + \frac{\Sigma \, fd \times \text{band width}}{\Sigma \, f} = 84.5 + \frac{-1 \times 10}{80} \quad \text{(band width here is 10)}$$

$$= 84.5 - 0.125 = 84.375$$

The approximate mean rateable value = £84.38

EXERCISE D

1. Using an assumed mean of 100, calculate the mean of the following numbers.
    105, 108, 100, 94, 96, 97, 89, 111.

2. Using an assumed mean height of 148 cms, calculate the mean of these heights (measured in cms)—
    140, 142, 151, 150, 148, 149, 146, 145.

3. These are the actual weights (to the nearest tenth of a gramme) of 12 small bags of sweets. Calculate the mean weight, using an assumed mean. Weights—

116.2 116.3 115.9 116   116.1 115.8

116   116.2 116.3 116.1 116.2 115.9

4. The number of ships arriving in Glasgow from overseas, each month in a recent year was as follows:

111 107 133 114 140 132 125 135

126 135 132 116

Calculate the mean monthly figure, using an assumed mean.

5. The heights of a class of boys, measured to the nearest cm, are shown below. Using an assumed mean, calculate the mean height.

Height (cm)	155	156	157	158	159	160	161	162	163	164	165	166	167
Frequency	1	1	5	3	3	3	5	2	4	0	1	1	1

6. The length of the right foot of each girl in a third year class was measured to the nearest half centimetre. Calculate the mean length of foot (using an assumed mean of 22.0 cm).

Length (cm)	20.0	20.5	21.0	21.5	22.0	22.5	23.0	23.5	24.0	24.5	25.0	25.5	26.0
Frequency	0	2	4	2	4	4	3	3	2	1	0	0	1

7. Repeat the last question, by measuring the right foot of all the girls *or* the boys in your own class. (Why should you not take the boys and girls together for this?)

8–11. Calculate the means for the questions in Exercise C using the assumed mean method.

The median from a frequency distribution

It is very easy to find the median of a set of raw scores by setting the scores out in numerical order and counting along till the middle is found.

The same method is used to find the median of a frequency distribution when the data is ungrouped.

Example
Find the median of the following scores.

Score	51	52	53	54	55	56
Frequency	3	4	7	6	3	1

Solution
There is a total frequency of 24, so the median must lie between the 12th and 13th scores. We examine the frequencies to see where the 12th and 13th scores lie. These scores are both 53, so the median must be 53.

When the data we are concerned with is grouped into class intervals, it is only possible from the distribution, to say that the median lies in a certain group. To find the median more exactly we draw the cumulative frequency curve for the distribution and estimate the median from this.

Example
Find the median of the marks gained by 50 pupils in an English examination.

Mark	21–30	31–40	41–50	51–60	61–70	71–80
Frequency	2	6	11	20	7	4

Solution
Construct a cumulative frequency table and draw the curve.

Mark	Cum. Freq.
21–30	2
21–40	8
21–50	19
21–60	39
21–70	46
21–80	50

Fig. 4.1 Cumulative Frequency Curve of Examination Marks

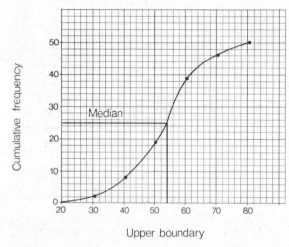

Fig. 4.2 (a) Geography Examination Results

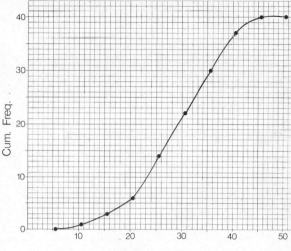

There is a total frequency of 50, so the median lies between the 25th and 26th score i.e. it is the $25\frac{1}{2}$th score.

We draw a line across the graph at the $25\frac{1}{2}$th score until it cuts the curve and then drop another line to the horizontal axis. Thus from the graph the median is the mark of 54.

Note: If the total frequency is N, then the

median is $\dfrac{N}{2} + \frac{1}{2}$

Where N is large the $\frac{1}{2}$ becomes insignificant and the median is taken as $\dfrac{N}{2}$.

EXERCISE E

1. The frequency curve (Figure 4.2a) shows the marks of 40 pupils in a geography exam.
    a) From the curve, estimate the median mark.
    b) If 25% of the pupils failed, what was the pass mark?

2. The curve in Figure 4.2b, shows the marks of the same class for a history exam.
    a) Estimate the median mark from the graph.
    b) If the pass mark was 24 what percentage of the class failed?
    c) What percentage of the class passed?

Fig. 4.2 (b) History Examination Results

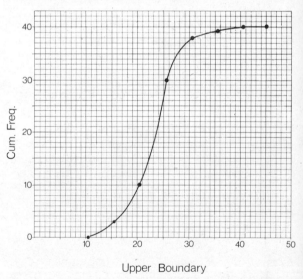

3–6. Find the cumulative frequency curves you drew for the questions in Exercise H, Chapter 3 and estimate the medians for each distribution.

7. Shown in the table on the right are the number of deaths registered in Wales for for each week of a recent year.

Construct a cumulative frequency table, draw the cumulative frequency curve and from it estimate the median number of deaths registered.

Number of Deaths	Frequency
450–499	3
500–549	16
550–599	16
600–649	11
650–699	3
700–749	2
750–799	1

Comparing the three 'averages'

The mean

Advantages
1. It can be calculated exactly.
2. It makes use of all the data.
3. It can be used in further statistical calculations.

Disadvantages
1. It can be very misleading if there is an abnormally high or low value, e.g. the ages of 10 pupils in a school are 5, 5, 5, 6, 6, 6, 7, 7, 7, 17 years. The mean age is 7.1 years. In this case 9 of the 10 pupils are below the mean age. The abnormally high value of 17 has unduly affected the mean.

The median

Advantages
1. It is simple to understand.
2. It is unaffected by abnormally high or low values.
3. It is characteristic of the normal group and sometimes represents an actual member of the group e.g. the pupil of median height in a class can actually be picked out and examined.

Disadvantages
1. It cannot be used in further statistical calculations.
2. Its value can only be estimated (from a cumulative frequency curve) in grouped distributions.
3. In small groups or in groups which have a rather odd pattern of distribution, it may not be characteristic of the group e.g. in a test there were 6 scores of 5 marks, 2 of 7 marks, 1 of 8 marks and 1 of 9 marks. From these marks the median would be 5 which would not be a good average to use for the group. (In this case, the mean would be a better average to use.)

The mode

Advantages
1. It is simple to understand.
2. It is unaffected by abnormally high or low values.
3. It is the average useful to manufacturers of shoes, clothes, hats etc.

Disadvantages
1. It cannot be determined exactly in a distribution where the data is grouped.
2. It cannot be used in arithmetical calculations.

45

5 APPLICATIONS OF THE MEAN

Moving averages

Consider the following information about the registration of new cars (in thousands) for the four quarters of 1969, 1970 and the first quarter of 1971.

New Registrations of Cars (in thousands)			
	1969	1970	1971
1st Quarter	233	250	272
2nd Quarter	253	275	
3rd Quarter	267	291	
4th Quarter	235	281	

These facts can be represented graphically as below.

The graph, like the data above it, is very irregular and shows principally the seasonal variations. It can be seen that there is a steady increase in registrations in the second and third quarters of each year and a sharp drop in registrations in the last quarter of the year. The data also shows that there is in general an increase in registrations for 1970 as compared with 1969 but the extent of this increase is difficult to assess from the data or the graph. An annual average would give an overall picture for each year but this would reduce the above list to two measures and there would be no way of using the odd measure for the first quarter of 1971.

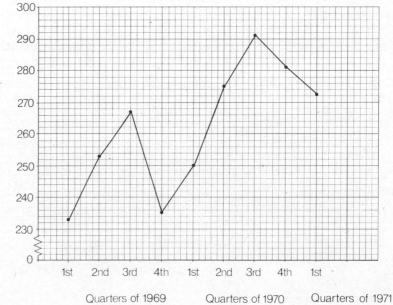

Registrations of cars in thousands

Fig. 5.1

This difficulty is common to many forms of measures which involve seasonal or other similar variations and some device which will smooth out the graph in order to show the general trend is required. One such device is what is commonly known as a moving average. In the above example the first average is the average for the four quarters of 1969. The next average is obtained by averaging the measures for the last three quarters of 1969 and the first quarter of 1970. The next average is obtained from the last two quarters of 1969 and the first two quarters of 1970. Further averages are obtained in succession by omitting the oldest measure and taking into account one new measure each time.

The necessary calculations for the moving averages can be tabulated as follows and it should be noted that the moving annual totals and the moving averages are listed opposite the middle point of the time interval concerned. Similarly in the graph of the moving averages the measure is plotted against the mid point of the time interval.

	Measure	Moving annual total	Moving averages
1969			
1st quarter	233		
2nd quarter	253		
		988	247
3rd quarter	267		
		1005	251
4th quarter	235		
		1027	257
1970			
1st quarter	250		
		1051	263
2nd quarter	275		
		1097	274
3rd quarter	291		
		1119	280
4th quarter	281		
1971			
1st quarter	272		

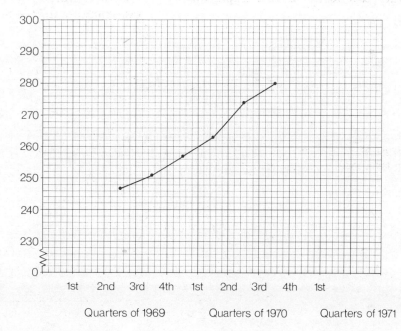

Registrations of car in thousands — Moving averages

Quarters of 1969   Quarters of 1970   Quarters of 1971

Fig. 5.2

It should be noted that the graph now shows a steady average increase in car registrations over the period of time concerned.

In this example since four measurements were averaged it is said to be a four point moving average, but if the data had been given in terms of monthly registrations then it would have been appropriate to average successive sets of 12 measures and this would be called a 12 point moving average. When seasonal variations are to be smoothed out then it is essential that sets of measures for complete seasons be taken for the calculation of the moving averages.

Moving averages can be used to smooth out a graph when the irregularities are the result of other than seasonal variations. The only change is that since there is no complete season available as the obvious size of group for the calculation of the averages an arbitrary choice must be made in the matter.

Example
The number of passes in higher mathematics in a Scottish Secondary School for each of the last 15 years are as follows:

Calculate the three-point moving averages for the above measures and illustrate the

Year	No. of Passes	Year	No. of Passes
1957	35	1965	44
1958	33	1966	48
1959	47	1967	68
1960	43	1968	56
1961	39	1969	65
1962	28	1970	72
1963	33	1971	74
1964	32		

raw measures and their moving averages on the same grid.

Solution

(Figure 5.3 and table opposite)

It can be noted that, in this case, the graph is only partially smoothed out by taking a three-point moving average. A more regular curve could be obtained by using, say, a five-point moving average.

In general moving averages are used to smooth out a graph or distribution of measures thereby eliminating the variations and bringing out the general trend of the measures. If the variations are seasonal then

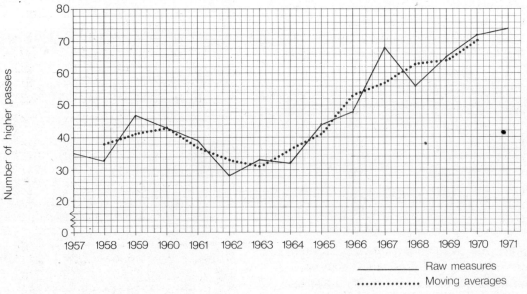

——————— Raw measures
.... Moving averages

Fig. 5.3

48

Year	No. of Passes	Three year Moving Totals	Three-point Moving Averages
1957	35		
1958	33	115	38
1959	47	123	41
1960	43	129	43
1961	39	110	37
1962	28	100	33
1963	33	93	31
1964	32	109	36
1965	44	124	41
1966	48	160	53
1967	68	172	57
1968	56	189	63
1969	65	193	64
1970	72	211	70
1971	74		

the group chosen for averaging should cover a complete season otherwise the size of group is an arbitrary choice.

EXERCISE A

1. Calculate the three-point moving averages of the following measures:

 20 15 5 10 12 8 15 25 10 15

2. Calculate the five-point moving averages of the measures quoted in question 1.

3. In a school in Scotland the number of passes in Higher French for each of the last 12 years was as follows:

Year	No. of Passes	Year	No. of Passes
1960	21	1966	26
1961	23	1967	32
1962	25	1968	23
1963	23	1969	31
1964	31	1970	39
1965	24	1971	35

Calculate the four-point moving averages for these measures and illustrate the actual number of passes and the moving averages on the same grid.

4. Consumer expenditure on cars and motor cycles, in millions of pounds, for each quarter of the years 1967 to 1970 were as listed below.

	1967	1968	1969	1970
1st quarter	150	153	155	183
2nd quarter	180	161	182	206
3rd quarter	216	189	200	219
4th quarter	228	209	187	237

Calculate the four-point moving averages for these measures and illustrate the raw data and the moving averages on the same grid.

5. The number of live births in England and Wales for the last two quarters of 1968 and for each quarter of the years 1969 and 1970 are listed below. Calculate the four-point moving averages for these measures and illustrate the raw measures and the moving averages on the same grid.

Live Births Registered (in thousands)			
	1968	1969	1970
1st quarter		213.1	196.3
2nd quarter		206.1	204.1
3rd quarter	207.7	199.0	196.5
4th quarter	190.3	180.0	187.6

6. The number of passenger movements by air and sea both into and out from the United Kingdom for each quarter of the years 1968 to 1970 are tabulated below.

1968	Inward		Outward	
	Sea	Air	Sea	Air
1st quarter	362	1058	352	1040
2nd quarter	1112	2119	1195	2146
3rd quarter	2643	3194	2577	3211
4th quarter	502	1451	496	1440
1969				
1st quarter	420	1237	419	1241
2nd quarter	1323	2527	1415	2534
3rd quarter	2932	3753	2912	3785
4th quarter	566	1728	563	1697
1970				
1st quarter	552	1494	589	1544
2nd quarter	1406	2929	1443	2891
3rd quarter	3144	4345	3103	4381
4th quarter	1533	2046	626	2029

Calculate four-point moving averages for:
  a) sea movements inward
  b) sea movements outward
  c) air movements inward
  d) air movements outward

7. The daily takings in a small gift shop for the five weeks prior to Christmas are as follows. Calculate the six-point moving averages for the takings and plot the graph of these moving averages. (Takings are given correct to the nearest pound.)

	Week 1 £	Week 2 £	Week 3 £	Week 4 £	Week 5 £
Mon.	36	40	46	45	45
Tue.	53	59	68	75	71
Wed.	46	39	35	50	39
Thur.	62	75	92	104	87
Fri.	85	94	138	152	121
Sat.	120	132	175	215	165

8. The total numbers of unemployed persons in the development areas, expressed as a percentage of the total population for the years 1958 to 1967 are listed below. Calculate the three-point moving averages for these measures.

Year	% unemployed	Year	% unemployed
1958	3.3	1963	4.2
1959	3.7	1964	3.1
1960	2.9	1965	2.4
1961	2.4	1966	2.2
1962	3.4	1967	3.5

9. The gross profits of companies as a percentage of their total income for the years 1959 to 1970 are as follows:

Year	% Profit	Year	% Profit
1959	15.1	1965	14.5
1960	15.5	1966	13.0
1961	14.3	1967	13.2
1962	13.7	1968	13.4
1963	14.7	1969	12.5
1964	15.1	1970	11.4

Calculate the four-point moving averages.

Weighted mean

In some situations it is unsatisfactory to calculate the ordinary mean of a set of measures. The situation may be such that greater emphasis must be given to particular measures. This is commonly done by using a weighted mean. The technique is best illustrated by example.

Example 1
To estimate the likely performance of a class in a national examination the pupils sat two separate tests which were both marked out of 100. The teacher considered that the second test was twice as valuable as the first. If a pupil scored 40% in the first test and 55% in the second then on this basis his weighted mean is obtained by averaging his first result with twice the second result.
i.e. Weighted Mean $=$
$$\bar{x}_w = \frac{40 + 2 \times 55}{1 + 2} = \frac{150}{3} = 50.$$

Example 2
If in the above the decision had been to weight the results in the ratio 2:3 then the Weighted Mean $=$
$$\bar{x}_w = \frac{2 \times 40 + 3 \times 55}{2 + 3} = \frac{245}{5} = 49.$$

Example 3
A manufacturer makes three models A, B and C on which he reckons to make a profit of £50, £40, £30 respectively. His results show that for every 2 models of type A which he sells he is able to sell 3 models of type B and 5 of type C. Find his mean profit per article sold.
Weighted Mean $=$
$$\bar{x}_w = \frac{2 \times £50 + 3 \times £40 + 5 \times £30}{2 + 3 + 5}$$
$$= \frac{£370}{10}$$
$$= £37$$

In general if the measures $x_1, x_2, x_3, \ldots x_n$ are given weights $w_1, w_2, w_3, \ldots w_n$ then their

Weighted Mean =

$$\bar{x}_w = \frac{w_1x_1 + w_2x_2 + w_3x_3 + \ldots + w_nx_n}{w_1 + w_2 + w_3 + \ldots + w_n}$$

EXERCISE B

1. Find the weighted mean of two measures 5 and 10 if they are given weights of 2 and 3 respectively.
2. Find the weighted mean of the measures 10, 14, 15, 18, 20 if they are given weights of 1, 2, 3, 4, 5 respectively.
3. In the three sections of an examination a candidate scored 40%, 45% and 60%. Find the weighted mean if the three sections are given weights 1, 2 and 2 respectively.
4. An alloy is made by mixing three parts, by volume of material A with five parts, by volume of material B. If the density of material A is 5 gm per cm^3 and that of material B is 9 gm per cm^3 find the density of the alloy.
5. The two isotopes of chlorine have atomic masses of 35 and 37. If the two isotopes are normally found in a mixture in the ratio of three of the first to one of the second find the average atomic mass of the mixture.
6. The two isotopes of copper have atomic masses of 63 and 65. If these isotopes are normally found in a mixture in the proportion of 7 of the first to 3 of the second find the average atomic mass of the mixture.

Index numbers

Two measures can be compared in a variety of ways:

1) the two measures can be stated
2) the difference between the measures can be given
3) the one measure can be given as a fraction of the other
4) the percentage change can be given
5) the one measure can be given as a percentage of the other.

The last of these methods of comparison is the concept underlying an index number. The only fundamental difference is that in an index number the percentage sign is omitted. Index numbers are at present normally used to express changes in prices and the figure upon which the percentage is calculated is called the base.

Example 1

An article which cost £5 in 1968 now costs £7. Find the index number corresponding to the present price if the 1968 price is taken as base.

$$\text{Index number} = \frac{7}{5} \times 100 = 140.$$

Example 2

In 1968 the index number corresponding to the cost of an article on the 1965 cost as base was 120. The present index number taken on the 1968 cost as base is 105. Find the index number for the present cost taken on the 1965 cost as base.

Index number = 105% of 120

$$= \frac{105}{100} \times 120 = 126$$

Example 3

If the index number for an article in 1968 was 120 and at present is 150, both index numbers being taken on the 1965 cost as base, find the index number for the present cost taken on the 1968 cost as base.

$$\text{Index number} = \frac{150}{120} \times 100 = 125.$$

EXERCISE C

1. Calculate the index numbers corresponding to the 2nd cost of each of the following commodities taking the first cost as base in each case:

Commodity	1st cost £	2nd cost £
A	2.00	2.10
B	5.50	6.60
C	2.50	2.00
D	34.50	53.20*
E	245	452*

2. In each of the following calculate the present cost if the given index number is the present one and is calculated on the first cost as base:

Commodity	1st cost £	Index number
A	10	105
B	7.80	120
C	0.84	125
D	345	97†
E	345	103†

3. In each of the following examples the 1970 and 1971 index numbers are calculated on the 1968 cost as base. Express the 1971 cost as an index number calculated on the 1970 cost as base.

Commodity	1970 Index No.	1971 Index No.
A	110	132
B	120	144
C	120	108
D	90	72
E	123	134*

4. The index number for an item in 1968 was 105, calculated on the 1966 cost as base, and in 1970 the index number for the same item calculated on the 1968 cost as base was 140. Calculate the index number for the 1970 cost on the 1966 figure as base.

*Correct to the nearest integer

†Correct to the nearest £

5. The index number of an article in 1970 was 90, calculated on the 1969 cost as base and the index number of the same article in 1971 was 110, calculated on the 1970 cost as base. Find the index number in 1971 for that article calculated on the 1969 cost as base.

Index of retail prices

This measure is frequently quoted in the press and changes in this measure are frequently used as grounds for claims for wage increases. Fundamentally this index is a combination of the ideas behind weighted mean and index numbers. A large representative sample of the population has been required to keep a very careful note of all forms of spending incurred by the family over a period of time and from the analysis of all of this information details of the spending of the average family have been obtained. Decisions have then been taken as to the classification of all of the items of spending and weights have been allocated to these classes of spending in proportion to the results obtained from the survey. The total of the weights allocated is normally 1000. The pattern of family spending changes as national habits change and thus the allocation of these weights is kept continually under review.

Some particular date is chosen as a base and each class of goods is allocated the index number 100 for that date. Changes in the cost of items within each class can be kept under review and appropriate index numbers can be obtained, at any time, for the various classes of expenditure. The weighted mean of these various index numbers can be calculated and this measure is called the index of retail prices. The full list of details of the index of retail prices is too long to be quoted here but a summary of the facts given as at June 1970 is as follows: (January 1962 = 100)

	Weight	Index
Food	298	142
Tobacco and Alcoholic drink	130	140
Housing, Fuel and Light	180	153
Household Goods	60	125
Clothing	86	123
Transport	126	131
Services	55	152
Miscellaneous	65	142
Total	1000	

Example 1

The cost of producing an article is weighted as follows: labour 55, materials 25, plant and maintenance and depreciation 20. On the 1969 costs as base the index numbers for these sections of the cost are 160, 135 and 110 respectively. Find the index number corresponding to the present cost of production on the basis of the 1969 costs.

Solution

	Weight	Index Number
Labour	55	160
Materials	25	135
Plant	20	110
Total	100	

Present Index Number
$$= \frac{55 \times 160 + 25 \times 135 + 20 \times 110}{55 + 25 + 20}$$
$$= \frac{14375}{100}$$
$$= 144$$

Example 2

If the 1968 index of retail prices had the following weights and indices as based on the 1962 prices find the 1968 index of retail prices.

Solution

	Weight	Index Number
Food	304	124
Tobacco and Alcohol	129	126
Housing and Durable Goods	331	127
Transport	120	119
Misc. Goods and Services	116	128
Total	1000	

Index of Retail Prices
$$= \frac{\begin{array}{c}304 \times 124 + 129 \times 126 + 331 \times 127 \\ + 120 \times 119 + 116 \times 128\end{array}}{304 + 129 + 331 + 120 + 116}$$
$$= \frac{125115}{1000}$$
$$= 125$$

EXERCISE D

1. A shopkeeper estimates his expenditure under the undernoted headings and allocates the accompanying weights and index numbers to the items taking his 1965 costs as base. Find the weighted index number corresponding to these details.

	Weight	Index Number
Wages	140	170
Cost of Goods	765	155
Rent and Rates	35	210
Maintenance of Premises	60	110

2. As part of the evidence for a claim for increased pocket money the boys of a boarding school produced the following details regarding the weighting of their spending and the changes in costs over the last year.

	Weight	Index Number
Sweets	25	125
Comics	10	130
Pictures	35	110
Toys etc.	20	120
Savings and presents	10	120

Find the weighted index number corresponding to these measures and state the percentage increase in pocket money that they would be justified in claiming on this evidence.

3. A housewife studied her outlay on food etc. and allocated weights to the various classes of expenditure as below. From a comparison of prices of the item involved she estimates the attached index numbers, using the 1969 prices as base. Calculate the weighted index number for the range of expenditure.

	Weight	Index Number
Groceries	42	105
Meat and fish	20	115
Fruit and vegetables	8	150
Milk and eggs	12	106
Bread and cakes	10	110
Cleaning materials	8	105

4. Calculate the weighted index number corresponding to the following items and their attached weights and index numbers.

	Weight	Index Number
Food	316	121
Tobacco and Drink	125	110
Housing	160	115
Clothing	83	109
Durable Goods	80	105
Transport	120	112
Services	60	115
Misc. Goods	56	110

5. Calculate the index number for all the groups listed below by using the given data.

	Weight	Index Number
Food	395	118
Tobacco and Drink	115	102
Rent and Rates	75	115
Fuel and Light	62	114
Clothing	83	98
Durable Goods	55	96
Transport	105	112
Misc. Goods and Services	110	100

MEASURES OF DISPERSION

Range

In Chapter 4, we studied measures of central tendency or averages. These averages are very important since they can give us a picture of the group which they represent. However, taken by themselves, they give us a very limited view of the whole picture. As well as the average, we need to know how the rest of the data is grouped around the average—whether it is closely gathered round or scattered more widely. In other words we have to find some way of measuring how the scores scatter. We need a measure of the spread, scattering or dispersion of the scores.

Examine the data shown below of the weekly wages of ten employees in different factories.

In each factory the mean wage is £30. But the way the wages are scattered about the mean is very different in each case. One simple way of measuring the scatter is to consider the **range** of values. In some situations this is helpful but in the above example, the range is the same in factory

A, B and C. As a measure of dispersion the range is very easily found but is not particularly good. Thus more satisfactory measures are required.

Semi-interquartile range

In many experimental situations we tend to distrust extreme measures and as a result these measures are often discarded. It is this type of thinking that leads to an improved version of the range called the **semi-interquartile range.**

Just as the median divides a set of scores into two equal parts, the **quartiles** divide a set of scores into four equal parts. There are three quartiles, the lower, the middle and the upper quartiles.

The lower quartile is called Q_1
The middle quartile (usually called the median) is Q_2
The upper quartile is Q_3
Then the inter-quartile range is $Q_3 - Q_1$
and the Semi-Interquartile Range is $\dfrac{Q_3 - Q_1}{2}$

Weekly Wage	Employees in Factory A	Employees in Factory B	Employees in Factory C	Employees in Factory D
£50	2	5	1	0
£40	2	0	2	4
£30	2	0	4	2
£20	2	0	2	4
£10	2	5	1	0

Example 1

Find the upper and lower quartiles and the semi-interquartile range of this set of scores.

8, 9, 7, 10, 5, 4, 6.

Solution

First write out the scores in numerical order.

4	5	6	7	8	9	10
	Q_1		Q_2		Q_3	

Fix the position of the median first, then since there are three scores below the median and three scores above the median the lower quartile is the second score and the upper quartile is the sixth score.

$$\text{Thus } Q_1 \text{ (lower quartile)} = 5$$
$$Q_3 \text{ (upper quartile)} = 9$$

and The Semi-interquartile Range $= \dfrac{9-5}{2} = 2$.

Example 2

Find the range and the semi-interquartile range of the following set of scores.

12, 12, 3, 8, 9, 9, 5, 6, 3, 4, 5, 11

Solution

First arrange the scores in numerical order.

3	3	4	5	5	6	8	9	9	11	12	12
			$Q_1 = 4.5$			$Q_2 = 7$			$Q_3 = 10$		

Then locate the median and since there are six scores below and above the median the lower quartile is the average of the third and fourth scores and the upper quartile is the average of the ninth and tenth scores.

$$\text{Thus the median} = \frac{6+8}{2} = 7$$

$$\text{The lower quartile} = \frac{4+5}{2} = 4.5$$

$$\text{The upper quartile} = \frac{9+11}{2} = 10$$

$$\text{The semi-interquartile range} = \frac{10-4.5}{2} = 2.75$$

$$\text{The range} = 12 - 3 = 9$$

The quartiles from a frequency distribution

We find the quartiles from a frequency distribution in the same way as we did for the median in chapter 4 i.e. we estimate the quartiles from the cumulative frequency curve of the distribution. When dealing with frequency distributions with a small number of measures we shall again consider the number of measures above and below the median as an aid to the location of quartiles.

Example

The cumulative frequency table for the marks gained by 50 pupils in an English examination is shown opposite. Since there are 50 measures there are 25 measures above and below the median and thus the lower quartile corresponds to the 13th score and the upper quartile corresponds to the 38th score.

Mark	Cum. Freq.
21–30	2
21–40	8
21–50	19
21–60	39
21–70	46
21–80	50

Figure 6.1 (a) shows the cumulative frequency curve for the distribution with the median and quartiles shown.

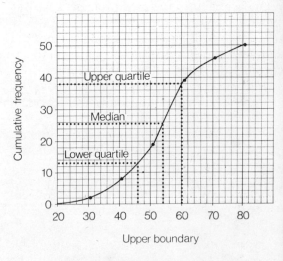

Fig. 6.1 (a) Cumulative Frequency Curve for the Distribution with the Median and Quartiles shown

From the graph,
the median   = 54
the lower quartile   = 46
the upper quartile   = 60
and thus the semi-interquartile range

$$= \frac{60 - 46}{2}$$

For large values of the total cumulative frequency, N, the fine details covered in the above discussion become rather insignificant and we can say that
the lower quartile corresponds to $\frac{1}{4}$N
the median corresponds to $\frac{1}{2}$N
the upper quartile corresponds to $\frac{3}{4}$N

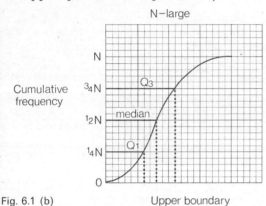

Fig. 6.1 (b)   Upper boundary

Inter-percentile range.
Just as the quartiles divide a distribution into four equal parts, the percentiles divide a distribution into one hundred equal parts.
   Obviously,
      the lower quartile is the 25th percentile,
      the median is the 50th percentile
and, the upper quartile is the 75th percentile.
   Again, as with the quartiles, we estimate the percentiles from a cumulative frequency curve.
   Using percentiles we have another measure of the dispersion of a distribution called the inter-percentile range. This is the range between the 10th percentile and the 90th percentile. Again, as with the inter-quartile range it does not depend on the extreme values, but as with the range, the inter-quartile and semi-interquartile ranges it does not lend itself to further mathematical treatment.

An easy way to obtain any percentile is to draw another axis to the right of the graph of the cumulative frequency curve and mark the scale from 0 to 100, making 100 level with the total frequency.
   If we want to find, say, the 10th and 90th percentile we draw lines across to the curve as shown below and so obtain the values of the variable.

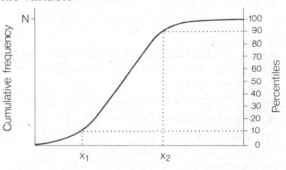

Fig. 6.2   Variable

EXERCISE A

1. Find the range and the semi-interquartile range of the following sets of scores.
   a) 20, 18, 17, 23, 19, 21.
   b) 11, 12, 13, 15, 7, 6, 5, 10, 14, 8, 9.
   c) 108, 108, 110, 101, 110, 111, 104, 104, 105, 107.
   d) 4, 5, 5, 1, 1, 2, 8, 8, 9, 9, 8, 7, 2.

2.

Weight (kg) (Upper Boundary)	Cumulative Frequency
66	5
67	8
68	12
69	20
70	35
71	47
72	59
73	68
74	72
75	75

This cumulative frequency table shows the weight in kg of 75 senior boys. Draw the

57

cumulative frequency curve and from it estimate:

  a) the median
  b) the semi-interquartile range
  c) the 10–90 percentile range
  d) how many boys weigh 69.5 kg or less
  e) how many boys weigh more than 71.5 kg.

3. 140 pupils sat two examinations in Arithmetic. The frequency table shows the distribution of their marks.

Marks	1st Exam. Freq.	2nd Exam. Freq.
0–10	0	4
11–20	3	6
21–30	10	9
31–40	34	9
41–50	42	12
51–60	32	20
61–70	12	42
71–80	5	22
81–90	2	10
91–100	0	6

Construct cumulative frequency tables and draw the cumulative frequency curves for each examination on the same graph.

  a) Estimate the medians
  b) Find the semi-interquartile ranges
  c) If the pass mark for the first examination was 45, how many pupils passed?
  d) What percentage of the pupils failed this examination?
  e) If 78 pupils failed in the second examination, what was the pass mark?
  f) In a few words compare the two distributions and comment on the standard of the test papers.

4. These are the heights (to the nearest cm) of some boys and girls of the same age.

Boys' heights

159	162	166	158	162	164	158	170
165	165	165	159	162	147	147	160
162	163	164	161	164	162	163	165
163	161	164	161	165	167	162	159
162	167	165	158	166	164	158	160
163	160	161	164	162	162		

Girls' heights

162	157	161	164	162	159	160	164
163	156	165	162	163	161	160	162
164	162	161	162	159	160	162	162
161	160	164	163	164	162	167	161
161	162	159	164	162	165	161	161
158	159	157	165	161	163		

Form a frequency table for the two distributions using groups starting at 147–148, 149–150 etc.

Construct a cumulative frequency table and draw the two curves on the same graph.

  a) Find the median for each distribution.
  b) Find the semi-interquartile ranges.
  c) How many girls are 159 cm or less in height?
  d) How many boys are 159 cm or less in height?
  e) How many girls are taller than 163 cm?
  f) How many boys are taller than 163 cm?
  g) Comment on the two distributions.

5. 60 pupils sat an examination in English and one in Arithmetic and the results are summarised below:

Mark	11–20	21–30	31–40	41–50	51–60	61–70	71–80	81–90
Frequency (English)	1	4	10	15	17	9	3	1
Frequency (Arithmetic)	0	1	2	20	32	4	1	0

By drawing cumulative frequency curves estimate the median, quartiles and the semi-interquartile range for each of the distributions.

Mean deviation

We require a measure of dispersion or variability which takes account of all the measures of the distribution. One such measure is the **mean deviation.** This, as it implies is the mean of the deviations or differences of the scores from either the mean, median or mode. It is usually most useful to calculate the mean deviation of the scores from the mean since the mean itself depends on all the measures in the distribution.

The mean deviation is calculated as shown below in the example. Since the negative signs must be ignored, any further algebraic development of the concept is made impossible, though the mean deviation would be quite satisfactory for many purposes.

Example

Calculate the mean deviation, from the mean, of the following scores.

10   12   14   16   18   20   22

Solution

$$\text{The mean, } \overline{X} = \frac{\sum X}{X} = \frac{112}{7} = 16$$

Score X	Deviations from mean $\overline{X} = 16$	Deviations taken as positive
10	−6	6
12	−4	4
14	−2	2
16	0	0
18	+2	2
20	+4	4
22	+6	6
		$\sum d = 24$

If the deviations were summed as they stand in the second column the answer would be 0, so all the deviations are taken as positive.

The sum of the deviations, $\sum d = 24$

$$\text{and the mean of the deviations} = \frac{\sum d}{N}$$

$$= \frac{24}{7}$$

$$= 3.43$$

EXERCISE B

Find the mean deviation, from the mean, of these sets of scores.
1. 1, 3, 4, 6, 7, 9, 12
2. 60, 63, 65, 66, 69, 70, 76
3. 1, 5, 11, 13, 15, 17, 20, 22

Standard deviation

If we add two extra steps to the calculation which was performed to find the mean deviation we obtain a measure of dispersion called the **standard deviation.** This is the most satisfactory measure of dispersion, since it makes use of all the scores in a distribution and is also quite acceptable mathematically.

In calculating the standard deviation, the mean of the distribution is found, the deviations of the scores from the mean are tabulated, these deviations are squared (thus dealing with the negative signs in an acceptable algebraic manner), and the mean of the squares of the deviations is calculated. The square root of this mean is the standard deviation.

Thus,

$$\text{the Standard Deviation} = \sqrt{\frac{\sum (X - \overline{X})^2}{N}},$$

where X stands for the various measures, $\overline{X}$ = the mean and N = the number of measures in the distribution.

Note: If we stop at finding the mean of the squares of the deviations, we have the **variance**

$$\text{i.e. the Variance} = \frac{\sum (X - \overline{X})^2}{N}$$

59

To calculate the standard deviation

To calculate the standard deviation of the scores 10, 11, 12, 13, 14, 15, 16 the following method is used.

1. List the scores in order

2. Calculate the mean $\bar{X} = \dfrac{\sum X}{N}$

$$= \dfrac{91}{7}$$

$$= 13$$

3. List the deviations of each score from the mean.

Score X	Deviation of score from mean $X - \bar{X}$	Squares of deviations $(X - \bar{X})^2$
10	$10 - 13 = -3$	9
11	$11 - 13 = -2$	4
12	$12 - 13 = -1$	1
13	$13 - 13 = \ \ 0$	0
14	$14 - 13 = +1$	1
15	$15 - 13 = +2$	4
16	$16 - 13 = +3$	9

$$\sum(X - \bar{X})^2 = 28$$

4. Square the deviations (this eliminates the negative signs).

5. Sum the squares of the deviations $\sum(X - \bar{X})^2 = 28$.

6. Divide this by the number of scores, N = 7.

7. Take the square root of the answer, which is the standard deviation

$$s \text{ (standard deviation)} = \sqrt{\dfrac{\sum(X - \bar{X})^2}{N}}$$

$$= \sqrt{\dfrac{28}{7}}$$

$$= \sqrt{4}$$

$$= 2$$

Example 1

Calculate the mean and standard deviation of the following scores.

22   24   26   28   30   32   34   36

Solution

Score (X)	$X - \bar{X}$	$(X - \bar{X})^2$
22	$22 - 29 = -7$	49
24	$24 - 29 = -5$	25
26	$26 - 29 = -3$	9
28	$28 - 29 = -1$	1
30	$30 - 29 = +1$	1
32	$32 - 29 = +3$	9
34	$34 - 29 = +5$	25
36	$36 - 29 = +7$	49

$$\sum(X - \bar{X})^2 = 168$$

$$\bar{X} = \dfrac{\sum X}{N}$$

$$= \dfrac{232}{8}$$

$$= 29$$

$$s = \sqrt{\dfrac{\sum(X - \bar{X})^2}{N}}$$

$$= \sqrt{\dfrac{168}{8}}$$

$$= \sqrt{21} \text{ (Use logarithms or square root tables)}$$

$$= 4.58$$

Mean = 29

Standard Deviation = 4.58

Example 2

Calculate the mean and standard deviation of this frequency distribution of test marks.

Mark	0	1	2	3	4	5	6	7	8	9	10
Frequency	1	2	3	5	6	7	4	3	2	1	0

Solution

With a frequency distribution we use exactly the same process as in example 1 but to the table we must add a frequency column, an fX column and an f$(X - \bar{X})^2$ column. The formula used now becomes

$$s = \sqrt{\dfrac{\sum f(X - \bar{X})^2}{\sum f}}$$

Mark (X)	f	fX	$(X-\bar{X})$	$(X-\bar{X})^2$	$f(X-\bar{X})^2$
0	1	0	$0-4.5 = -4.5$	20.25	20.25
1	2	2	$1-4.5 = -3.5$	12.25	24.50
2	3	6	$2-4.5 = -2.5$	6.25	18.75
3	5	15	$3-4.5 = -1.5$	2.25	11.25
4	6	24	$4-4.5 = -0.5$	0.25	1.50
5	7	35	$5-4.5 = +0.5$	0.25	1.75
6	4	24	$6-4.5 = +1.5$	2.25	9.00
7	3	21	$7-4.5 = +2.5$	6.25	18.75
8	2	16	$8-4.5 = +3.5$	12.25	24.50
9	1	9	$9-4.5 = +4.5$	20.25	20.25
10	0	0	$10-4.5 = +5.5$	30.25	0
	$\sum f = 34$	$\sum fX = 152$			$\sum f(X-X)^2 = 150.50$

$$\bar{X} = \frac{\sum fX}{\sum f}$$

$$= \frac{152}{34}$$

$$= 4.5 \text{ (correct to 1st decimal place)}$$

$$s = \sqrt{\frac{\sum f(X-\bar{X})^2}{\sum f}}$$

$$s = \sqrt{\frac{\sum f(X-\bar{X})^2}{\sum f}}$$

$$= \sqrt{\frac{150.5}{34}}$$

$$= \sqrt{4.43} \text{ (correct to 2nd decimal place)}$$

$$= 2.10$$

Mean = 4.5

Standard deviation = 2.10

Note: When working with a grouped distribution use the mid-point of the interval to stand for the group and proceed as above.

EXERCISE C

1. Calculate the mean and standard deviation of each of the following sets of scores.
   a) 3, 4, 5, 6, 7, 9, 10, 12.
   b) 24, 29, 27, 20, 20, 23, 21.
   c) 66, 66, 60, 64, 69, 67, 63.
   d) 35, 37, 37, 40, 30, 34, 33, 38.
   e) 30, 33, 24, 28, 20, 17, 25, 39, 34, 42.

2. In ten successive years in Glasgow the highest temperature (in the shade) for each year was recorded as:

   78, 82, 82, 80, 79, 76, 76, 78, 72, 74. (°F)

   Calculate the mean and standard deviation of these temperatures.

3. In the same ten years, the lowest temperature (in the shade) recorded each year was:

   12, 24, 15, 18, 12, 15, 18, 11, 19, 11. (°F)

   Calculate the mean and standard deviation of the temperatures.

4. Below are the heights of 40, fourth year boys, measured to the nearest 2 cm. Calculate the mean height and the standard deviation of the heights.

Height (cm)	150	152	154	156	158	160	162	164	166	168	170	172	174	176	178	180
Frequency	1	1	2	3	4	4	5	6	4	4	1	2	2	0	0	1

5. These are the heights of 30 fourth year girls measured at the same time as the boys.

Height (cm)	146	148	150	152	154	156	158	160	162	164	166	168	170
Frequency	1	2	2	3	3	7	2	5	1	2	1	0	1

Calculate the mean and standard deviation of the girls' heights and compare them to your answers for question 4. Give reasons for any differences you find.

6. Measure the heights of a number of boys or girls in your school and calculate the mean and standard deviation.

7. The number of days in which rain fell (rain days) each month in 1964 were recorded as follows.

Month	North Scotland	East Scotland	West Scotland
Jan.	23	18	22
Feb.	19	15	18
Mar.	19	16	17
Apr.	19	16	17
May	17	15	16
June	17	14	16
July	19	17	18
Aug.	19	17	19
Sept.	21	16	19
Oct.	22	19	21
Nov.	22	18	20
Dec.	22	18	21

For each region calculate the mean and standard deviation of the number of rain days.
  Compare briefly the three distributions explaining any differences or similarities.

8. The frequency table below shows the distribution of marks in an examination.
  Calculate the mean and standard deviation of the marks.

11. This table shows the death rates per 1000 population of a number of countries for 1964.

Form a frequency table grouping the data, starting at 6.0—6.9, 7.0—7.9 etc. Calculate the mean and standard deviation of the death-rates for these countries.

Country	Death-rate per 1000
Eastern Germany	13.5
Austria	12.3
Belgium	12.1
Scotland	11.7
Eire	11.3
England and Wales	11.0
Western Germany	10.7
France	10.5
Northern Ireland	10.2
Portugal	10.0
Denmark	10.0
Norway	10.0
Sweden	10.0
Hungary	9.9
Czechoslovakia	9.6
Italy	9.6
U.S.A.	9.4
Yugoslavia	9.4
Finland	9.3
Switzerland	9.2
Australia	9.0
New Zealand*	8.8
Spain	8.7
Netherlands	7.7
Canada	7.6
Poland	7.6
Japan	6.9
Israel†	6.0

* excluding Maoris
† Jewish population only.

Mark	1–5	6–10	11–15	16–20	21–25	26–30	31–35	36–40	41–45	46–50
Frequency	0	2	4	11	20	10	5	3	2	1

9. The weights of 40 fourth year boys measured to the nearest kilogramme, are shown below. Calculate the mean and standard deviation of the weights.

Weight (kg)	35–39	40–44	45–49	50–54	55–59	60–64	65–69	70–74	75–79	80–84
Frequency	1	6	9	9	6	4	3	0	1	1

10. Take the weights of 40 fourth year boys in your school and calculate the mean and standard deviation. Compare your results with those of question 9.

Calculation of the standard deviation from an assumed mean

The calculation of the standard deviation as discussed in the last section is reasonable as long as the mean of the distribution is a whole number but in most distributions this is not the case and then there is a choice between accuracy and excessively heavy arithmetical calculations. To overcome this difficulty we look to the calculation of the standard deviation by using an assumed mean.

If $\bar{X}$ is the mean of the distribution and X_0 is the assumed mean then

$$\sum(X-\bar{X})^2$$

$$= \sum(X-X_0+X_0-\bar{X})^2$$

$$= \sum[(X-X_0)+(X_0-\bar{X})]^2$$

$$= \sum(X-X_0)^2+\sum 2(X-X_0)(X_0-\bar{X}) + \sum(X_0-\bar{X})^2$$

$$= \sum(X-X_0)^2+2(X_0-\bar{X})\sum(X-X_0) + N(X_0-\bar{X})^2$$

$$= \sum(X-X_0)^2+2(X_0-\bar{X})(\sum X-NX_0) + N(X_0-\bar{X})^2$$

$$= \sum(X-X_0)^2+2(X_0-\bar{X})(N\bar{X}-NX_0) + N(X_0-\bar{X})^2$$

$$= \sum(X-X_0)^2-2N(X_0-\bar{X})^2+N(X_0-\bar{X})^2$$

$$= \sum(X-X_0)^2-N(X_0-\bar{X})^2$$

Thus Standard Deviation

$$= \sqrt{\frac{\sum(X-\bar{X})^2}{N}}$$

$$= \sqrt{\frac{\sum(X-X_0)^2-N(X_0-\bar{X})^2}{N}}$$

$$= \sqrt{\frac{\sum(X-X_0)^2}{N}-(X_0-\bar{X})^2}$$

If d measures the deviates of the measures from the assumed mean then

$$\text{Standard Deviation} = \sqrt{\frac{\sum d^2}{N}-\left(\frac{\sum d}{N}\right)^2}$$

When this formula is extended to cover a frequency distribution it becomes:

$$\text{Standard Deviation} = \sqrt{\frac{\sum fd^2}{\sum f}-\left(\frac{\sum fd}{\sum f}\right)^2}$$

Finally if we have a grouped frequency distribution and we use the group width as the unit of measure of deviation from the assumed mean then the last formula would require to be multiplied by the group width. Thus if C is the group width

$$\text{Standard Deviation} = C\sqrt{\frac{\sum fd^2}{\sum f}-\left(\frac{\sum fd}{\sum f}\right)^2}$$

Example 1
Calculate the mean and standard deviation of
the distribution of heights below:

Height (cm)	162	163	164	165	166	167	168	169	170	171	172
Frequency	1	3	9	13	18	20	12	11	7	4	2

Solution

X	f	X_0	d	fd	fd^2
162	1		−5	−5	25
163	3		−4	−12	48
164	9		−3	−27	81
165	13		−2	−26	52
166	18		−1	−18	18
				−88	
167	20	167	0	0	0
168	12		1	+12	12
169	11		2	+22	44
170	7		3	+21	63
171	4		4	+16	64
172	2		5	+10	50
				+81	$\sum fd^2 = 457$
				$\sum fd = -7$	

$$\text{Mean} = 167 + \frac{(-7)}{100} = 166.9 \text{ cm (to 4 sig. fig.)}$$

$$\text{Standard Deviation} = \sqrt{\frac{\sum fd^2}{\sum f} - \left(\frac{\sum fd}{\sum f}\right)^2}$$

$$= \sqrt{\frac{457}{100} - \left(\frac{-7}{100}\right)^2}$$

$$= \sqrt{4.57 - 0.0049}$$

$$= \sqrt{4.565} = 2.13 \text{ cm}$$

Example 2
Calculate the mean and standard deviation
of the following distribution of class marks.

Mark	10–19	20–29	30–39	40–49	50–59	60–69	70–79	80–89
Frequency	1	5	7	9	18	14	4	2

Solution

X	f	X_0	d	fd	fd^2
10–19	1		−4	−4	16
20–29	5		−3	−15	45
30–39	7		−2	−14	28
40–49	9		−1	− 9	9
				−42	
50–59	18	54.5	0	0	0
60–69	14		+1	+14	14
70–79	4		+2	+8	16
80–89	2		+3	+6	18
	$\sum f = 60$			+28	$\sum fd^2 = 146$
				$\sum fd = -14$	

(Note that the unit of measure of d is the
group width)

$$\text{Mean} = 54.5 + \frac{(-14) \times 10}{60}$$

$$= 54.5 - 2.3 = 52.2$$

$$\text{Standard Deviation} = C\sqrt{\frac{\sum fd^2}{\sum f} - \left(\frac{\sum fd}{\sum f}\right)^2}$$

$$= 10\sqrt{\frac{146}{60} - \left(\frac{14}{60}\right)^2}$$

$$= 10\sqrt{2.433 - 0.054}$$

$$= 10\sqrt{2.378}$$

$$= 10 \times 1.54$$

$$= 15.4$$

Calculation of standard deviation using calculating machine

The calculation of the standard deviation
from the actual mean can be made less
tedious if an adding machine is used as the
deviates can be squared in turn and the sum
of squares can be accumulated in the product
register. It is when we use the assumed mean
method that the machine really comes into
its own as if we choose zero as our assumed
mean then the basic formula reduces to:

$$\text{Standard Deviation} = \sqrt{\frac{\sum X^2}{N} - \left(\frac{\sum X}{N}\right)^2}$$

and if we set the number 1 at the front of the setting register and each of the X's in turn at the back of the register, multiply each X by itself and allow the results to accumulate in the product register we get $\sum X$ at the front and $\sum X^2$ at the back of the product register. When dealing with individual results all tabulation is thus made unnecessary and the value of the standard deviation can be written down from the above formula.

5. Calculate the mean and standard deviation of each of the distributions of marks tabulated in Chapter 6, Exercise A, Question 3.

6. Calculate the mean and standard deviation of each of the distributions of heights tabulated in Chapter 6, Exercise A, Question 4.

7. Calculate the mean and standard deviation of each of the distributions of marks tabulated in Chapter 6, Exercise A, Question 5.

EXERCISE D

1. The heights of 26 pupils, measured to the nearest cm, are tabulated below. Find the mean and standard deviation of the distribution.

Height (cm)	148	149	150	151	152	153	154	155	156
Frequency	1	1	2	3	6	7	3	2	1

2. Find the mean and standard deviation of the following distribution of scores:

Score	51	52	53	54	55	56
Frequency	3	4	7	6	3	1

3. The rateable values of eighty houses were found, to the nearest pound, and the results are shown in the following table. Find the mean and standard deviation of the distribution.

Rateable value in pounds	50–59	60–69	70–79	80–89	90–99	100–109	110–119
Frequency	2	10	20	24	12	11	1

4. Calculate the mean and standard deviation of the following distribution of class marks.

Mark	1–5	6–10	11–15	16–20	21–25	26–30	31–35	36–40	41–45
Frequency	1	3	5	7	13	9	7	3	2

66

Standard scores

The comparison of measures from different distributions is usually unsatisfactory and the addition of such results frequently produces unexpected results. These difficulties are particularly obvious in the comparison of school examination results and when we add these marks to produce an average performance. The surprising results are usually caused by an unusual mean or dispersion in one or more of the distributions. These difficulties can be overcome if we use what are commonly called **standard** or **standardised scores**. When using standard scores we wish to eliminate the effect of the mean and the dispersion and the first of these is done by subtracting the mean of the distribution from each score. The deviate so obtained is then divided by the standard deviation to remove the effect of the dispersion. The result of this division is called the standard score. Standard scores from different distributions can then be compared or combined without being affected by either bias.

Example 1
A boy scored 50 marks in an English examination where the class mean was 45 and the standard deviation 5. He scored 60 marks in an Arithmetic examination where the class mean was 55 and the standard deviation 10. Compare his standing in the two subjects.

Solution

Standard Score in English $= \dfrac{50-45}{5} = \dfrac{5}{5} = +1$

Standard Score in Arithmetic $= \dfrac{60-55}{10} = \dfrac{5}{10} = +\tfrac{1}{2}$

Thus the boy did better in English than Arithmetic, compared to the rest of the class.

Example 2
Tom, Dick and Harry sat examinations in Maths, English and French. Their scores, the class means and standard deviations for each subject are shown in the following table.

	Maths	English	French	Total
Tom	81	50	46	177
Dick	42	57	55	154
Harry	33	56	65	154
Class Mean	48	52	50	
Standard Deviation	15	4	10	

Using the given raw scores Tom would be placed first. By using standard scores check the validity of this placing.

Solution

	Standard Scores			
	Maths	English	French	Total
Tom	$\dfrac{81-48}{15} = +2.2$	$\dfrac{50-52}{4} = -0.5$	$\dfrac{46-50}{10} = -0.4$	$+1.3$
Dick	$\dfrac{42-48}{15} = -0.4$	$\dfrac{57-52}{4} = +1.25$	$\dfrac{55-50}{10} = +0.5$	$+1.35$
Harry	$\dfrac{33-48}{15} = -1.0$	$\dfrac{56-52}{4} = +1.0$	$\dfrac{65-50}{10} = +1.5$	$+1.5$

The correct order of merit is Harry, Dick, Tom.

EXERCISE E

1. Jane scored 66 marks in her Geography examination in November where the mean was 60 and the standard deviation 6. In May, when the mean was 40 and the standard deviation 10, she had a mark of 55 for her Geography. In which examination did she show her ability better?

2. The following table shows the marks of a boy in three successive mathematics examinations.

Exam.	Mark	Mean	Standard Deviation
1st exam.	50	53	3
2nd exam.	51	58	7
3rd exam.	53	48	5

    a) In which examination was his relative standing best?

    b) Was his performance better in the second examination than in the first?

3 a). In an Arithmetic test given to a class, the mean and standard deviation were 64 and 8 respectively. The table below shows the marks of some of the pupils. Complete the table.

Pupil	Mark	Deviate from mean	Standard Score
Alan	68	$68 - 64 = 4$	$+\frac{1}{2}$
Betty	72	$72 - 64 = 8$	$+1$
Colin	56		
David	60		
Evelyn	80		
Fay	76		

b). In another Arithmetic test given to the same class a few weeks later, the mean was 49 and the standard deviation 6. Complete the following table.

Pupil	Mark	Deviate from mean	Standard Score
Alan	58		
Betty	52		
Colin	43		
David	49		
Evelyn	64		
Fay	55		

c). i) Which of the six pupils improved most?

    ii) Which pupil came down most in the second test?

    iii) Which pupil was most consistent in the two tests?

4. In the first term John scored 94% in Maths and Mary scored 80% while in the second term John scored 78% and Mary scored 90%. If the mean scores were first term: 50%, and second term: 60%, and the standard deviations were first term: 20%, and second term: 15%, find which pupil had the better average performance.

5. Mary, Jean and Helen sat examinations in English, French and Latin. The results and the class means and standard deviations are tabulated below. Find the correct order of merit.

	English	French	Latin
Mary	59	73	79
Jean	58	81	75
Helen	64	74	74
Class Mean	52	60	65
Standard deviation	5	10	8

7 REGRESSION AND CORRELATION

Regression and correlation

In this study, to date, there has only been one form of measure under consideration at any one time. In this chapter the interest is in situations where two forms of measure are being considered at the same time e.g. the height and weight of a group of children, attainment in arithmetic and the intelligence quotient of a group of 10 year old children. The incentive to compare measures in pairs as suggested above usually arises from a belief that the measures are in some way linked. The aim of this chapter is to find a method of illustrating the measures so that their interdependence, if it exists, can be seen, to measure the amount of their interdependence and use the given data to estimate other results concerning the given situation.

Scatter diagrams

The standard method of illustrating this type of situation is by using a scatter diagram and this consists of laying out each of the measures along one of the axes of a grid, then taking each item, in turn, the two measures for that item act exactly like an ordered pair and thus like the coordinates of a point on the grid. Each item considered is thereby linked to one point on the grid and that point can be plotted in the normal way and is usually marked on the grid as an 'X'. The points corresponding to all of the items being measured are similarly located on the grid and plotted and thus a scatter of 'Xs' is built up on the grid.

Example

The reading speeds in words per minute and the intelligence quotients of a group of 12 children were measured and are tabulated below.

The scatter diagram (Figure 7.1) is built up as follows. The axes are set out to cover a suitable range of measures of Reading Speed and Intelligence Quotient. Each child is then considered in turn and the two measures e.g. for the child A, Reading Speed 120 and Intelligence Quotient 80, locate one point on the grid and this point is marked with a dot. Similar dots are entered in turn for the other children and the complete scatter diagram will consist of 12 dots.

Child	A	B	C	D	E	F	G	H	I	J	K	L
Reading Speed. in words per minute	120	140	100	170	130	190	220	140	180	240	200	270
Intelligence Quotient	80	90	90	100	100	105	110	110	115	120	125	130

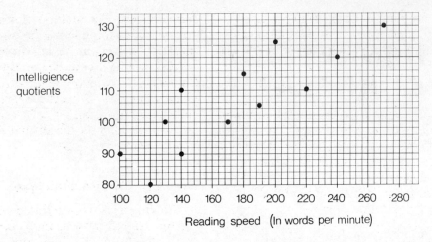

Intelligience quotients

Reading speed (In words per minute)

Fig. 7.1 Reading Speed (In Words per Minute)

From an examination of the pattern of crosses obtained it can be accepted that the above results help to support the belief that there is some linkage between Reading Speed and Intelligence Quotient but since there is some scattering of the points plotted indications are that the two measures are not very closely linked. It must also be accepted that no conclusive results could ever be obtained from such a small sample of the population.

The distribution of points plotted on a scatter diagram can give an indication of the relation which exists between the two characteristics being measured. Three basic forms of scatter diagram are shown below.

In Figure 7.2a the points plotted are scattered at random over the grid and this indicates that the measures are not related. Figure 7.2b shows the other extreme position as here the points are scattered along a straight line and this is the typical diagram for measures which are directly proportional to each other. In a scatter diagram like Figure 7.2c the closer the points are to being on a straight line the greater the degree of their possible linkage.

For measures to be related it is not necessary that the scatter diagram approximates to a straight line through the origin or even to any straight line. If the points plotted on the scatter diagram fit some straight line or curve this can be taken

Fig. 7.2

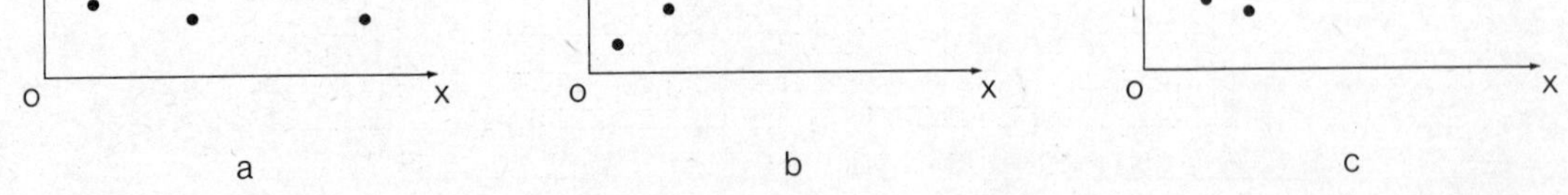

a

b

c

as evidence for some relation between the measures and it then becomes necessary to put the relation into algebraic form.

Regression

The results of scientific experiments can often be illustrated by the use of scatter diagrams and if these results indicate that the measures under consideration are related it is then desirable that the results are interpreted in a mathematical form. If the points plotted are almost on a straight line then the line which most nearly fits all of the points must be drawn. This line is called the 'Best Fitting Line' or the 'Regression Line'. If the measures are such that the points plotted are almost exactly on a straight line then there are few complications about the drawing of the line which most nearly fits all of the points but if the points are more scattered it then becomes possible to draw two regression lines. If the variables are, for convenience, called x and y then the one line is called the regression line of x on y and in it the values of y are assumed to be accurate and the points deviate from the straight line because of errors in the measurement of x. These deviations in the values of x will hereafter be called offsets. The method of drawing such a regression line will be discussed in the next section but having drawn this regression line it could be used to estimate the value of x which would correspond to a given value of y. The other regression line, called the regression line of y on x is drawn in a similar fashion except that the roles of x and y are interchanged.

Drawing of regression lines

In drawing a regression line by inspection the line chosen must as far as possible satisfy the following three conditions:
1) The line should pass through the point whose coordinates are the means of the two separate distributions of measurements i.e. the point $(\bar{x}, \bar{y})$.
2) The offsets on the one side of the line should balance those on the other side of the line.
3) The sum of all offsets should be as small as possible.

In conditions 2) and 3) it is really the squares of the offsets that should be considered but this is usually too difficult when the task is being carried out visually and it is normally quite satisfactory to use the above simplification.

Fig. 7.3 Regression Line of y on x

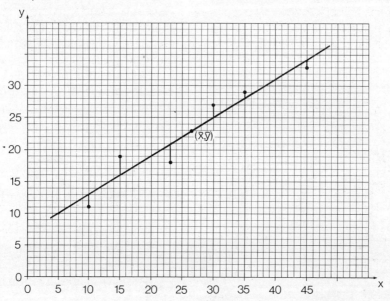

Example 1

Figure 7.3 shows a fairly typical drawing of a regression line. For this line the given data was as follows:

x	10	15	23	30	35	45
y	11	19	18	27	29	33

From these measures the coordinates of the point $(\bar{x}, \bar{y})$ can be found to be $(26\frac{1}{3}, 22\frac{5}{6})$. The offsets are shown as the vertical lines joining the points which were plotted with the corresponding points on the line which share the same x coordinate. Note that since in this case it is assumed that the values of x are accurate the offsets are all parallel to the y axis and measure the deviations of the points from the line in the direction of the y coordinate only. If the regression line of x on y is required the offsets are then the deviations of the points from the chosen line in the direction of the x axis and would then be drawn parallel to the x axis.

Equation of the regression line

Once the regression line has been drawn it is usually necessary that the equation of the line be expressed in mathematical terms. Consider the line PQ shown on the following diagram.

Fig. 7.4

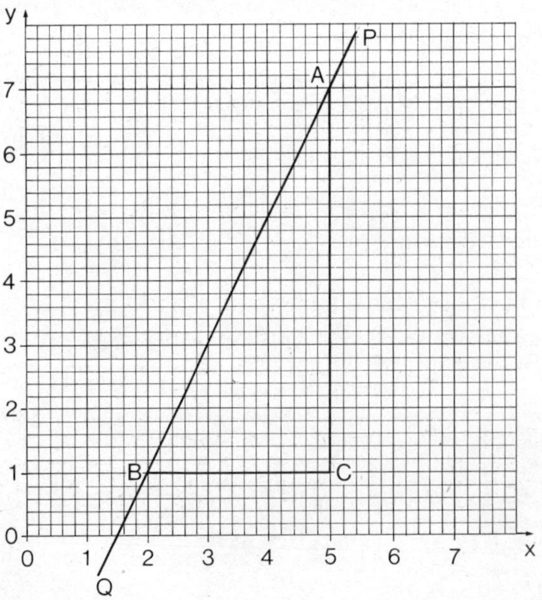

On the line select two points A and B which are well separated, one near each end of the line. From these points A and B draw lines AC and BC parallel to the axes in order to complete the right angled triangle ABC. The standard form of the equation of a straight line is $y = mx + c$ where m is the gradient of the line and c is a constant. If the coordinates of the points A and B are represented by the symbols $A(x_A, y_A)$ and $B(x_B, y_B)$ then the gradient of $AB = m$ is given by the fraction $\dfrac{CA}{BC} = \dfrac{y_A - y_B}{x_A - x_B}$. In the context of regression this fraction also gets the name of the **coefficient of regression.** Once the value of m has been established the constant c can be obtained by the substitution of the coordinates of either A or B into the equation $y = mx + c$. A good choice of the points A and B can help to reduce the weight of calculation. The points should be chosen so that the coordinates can be read without the use of too many decimals or fractions and since the calculation involves division by BC it is desirable that the length BC be chosen so as to make for an easy division.

Example 2

Consider the regression line drawn in Example 1 and find the equation of the regression line. Also use this equation to estimate the value of y which would correspond to 1) x = 20   2) x = 55.

Select the points A and B as A (45,34) and B (5,10) and let the equation of the line be of the form $y = mx + c$, then

$$m = \frac{y_A - y_B}{x_A - x_B} = \frac{34 - 10}{45 - 5} = \frac{24}{40} = \frac{3}{5}$$

The value of c can now be found using $m = \frac{3}{5}$ along with the coordinates of B, i.e.

$$10 = \frac{3}{5} \times 5 + c \quad \therefore \quad \underline{c = 7}$$

$$\therefore \text{ equation of regression line is } y = \frac{3}{5}x + 7$$

When x = 20, y = $\frac{3}{5} \times 20 + 7 = 12 + 7 = 19$

When x = 55, y = $\frac{3}{5} \times 55 + 7 = 33 + 7 = 40$

EXERCISE A

1. Find the equation of the regression line which passes through:
a) (7,8) and (4,2)   b) (2,5) and (5,2)
c) (5,6) and (21,14) d) (2,3) and (8,7)

2. The grid shown below shows two regression lines. Find the equation of a) AB b) CD.

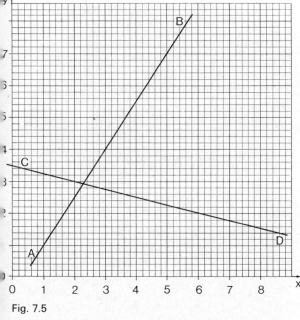

Fig. 7.5

3. Draw the scatter diagram corresponding to the following data.

x	3	6	11	13	20	25	30	37
y	6	9	10	12	14	18	21	23

On the diagram draw the best fitting line and find its equation. Use the equation to estimate the value of y which corresponds to x = 50.

x	86	90	93	94	95	96	98	100	100	105	110
y	120	104	100	80	70	38	98	25	20	50	10

4. Draw the scatter diagram and the best fitting line for the following set of measures of x and y.

x	2.2	2.8	3.2	4.0	4.2	5.4	6.0
y	7.8	6.8	6.0	4.5	3.8	1.8	1.2

Find the equation of the best fitting line and use that equation to find 1) the value of y when x = 1   2) the value of x when y = 0.

5. The marks obtained by 8 students in a formal examination and in an objective test in mathematics are tabulated below. The marks listed under the symbol x are those for the formal test and under y are the marks in the objective test.

x	5	9	14	15	25	27	32	33
y	10	11	12	15	15	17	20	20

Illustrate these measures on a scatter diagram, draw the regression line of y on x and find its equation.

6. Draw a scatter diagram to illustrate the following measures. On the diagram draw the regression line of y on x and find its equation.

x	5	9	15	15	23	23	30	32
y	10	2	10	19	15	22	26	32

7. Draw a scatter diagram to illustrate the distribution below.

On the scatter diagram draw and find the equation of a) the regression line of y on x b) the regression line of x on y.

8. A body is dropped from a helicopter and its height above ground level is estimated at various times after release. The following table gives the results of these observations where y measures the height above ground level in metres and t measures the time after release in seconds.

t	2	3	4	5	6	7	8
y	380	360	320	270	220	160	80

Since it is known that the height should be given by a formula of the type $y = mt^2 + c$ draw up a table showing the mapping of y against t^2 and illustrate this mapping on a scatter diagram. Draw the best fitting line for this diagram and use it to find the values of m and c. Find 1) the height from which the body was dropped i.e. the value when $t = 0$, 2) the time taken by the body to reach the ground i.e. the value of t when $y = 0$, 3) the time taken by the body to fall halfway to the ground.

9. The cost of producing lace decreases as the amount of lace produced increases and some measures relating to this fact are illustrated in the following table where x is the length of lace produced in hundreds of metres and y is the cost of the lace in pence per metre.

x	$\frac{1}{4}$	$\frac{1}{3}$	$\frac{1}{2}$	2	3	6
y	29	22	15	7.5	5.5	5

Since it is known that x and y are connected by an equation of the form $y = \frac{m}{x} + c$ draw up a table showing the corresponding values of $\frac{1}{x}$ and y and plot these results on a scatter diagram. Draw the best fitting line for this diagram and use it to find the values of m and c. Use the equation to find a) the cost per metre of lace if 1000 metres are produced   b) the number of metres of lace that must be produced in order that the cost per metre is 4.5 pence.

Correlation

While regression measures the line which best fits a certain set of measures it takes no account of how closely these measures approximate to that line. In the study of correlation the interest is in measuring the degree by which two distributions match each other. The type of measure that is used for this purpose is called the **coefficient of correlation** and it is assessed on a scale which runs from $+1$ through zero to -1. A coefficient of correlation of $+1$ means that the two distributions match each other perfectly and this would correspond to a scatter diagram like that shown in Figure 7.6 below where all of the points plotted lie along the leading diagonal of the grid. A coefficient of correlation of -1 would correspond to a pair of distributions where the measures are in completely opposite order i.e. the first of the one distribution is last in the other, the second in the first distribution is second last in the other distribution and so on. This type of situation would correspond to a scatter diagram like that shown in Figure 7.7 below where all of

Fig. 7.6

Fig. 7.7

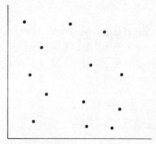

Fig. 7.8

the points plotted lie along the other diagonal of the grid. A zero coefficient of correlation would correspond to a pair of distributions which were completely unconnected and would have a scatter diagram like Figure 7.8 (page 74) where the entries are scattered at random over the grid. Intermediate values of the coefficient of correlation correspond to distributions whose scatter diagrams are intermediate between these extreme cases.

There are fundamentally two methods of assessing a coefficient of correlation, the one depends on the actual measures in the two separate distributions being considered and is called the Product-Moment Coefficient of Correlation and the other which is called the Coefficient of Rank Correlation depends only on the order of merit of the measures in the distributions. The first of these methods involves a considerable amount of heavy calculation and thus in this study only the second of these methods of arriving at a measure of correlation will be used.

Coefficient of rank correlation

The theory behind this measure is beyond the scope of the present study and thus only the procedure will now be discussed. Each distribution must first be put into an order of merit and traditionally this is taken in a descending order. Each item being considered has now two ranks allocated to it and the difference between these two ranks can now be found and if the symbol d is used to represent this difference then:

Coefficient of Rank Correlation

$$= R = 1 - \frac{6\sum d^2}{n(n^2 - 1)}$$

where n is the number of items in the distributions.

If two or more measures in one distribution are equal it is convenient, though not mathematically justifiable, to allocate to them a rank which is the average of the ranks which they would have occupied if they had been different e.g. if the 3rd and 4th measures in a distribution are equal they would both be allocated the rank $3\frac{1}{2}$ or if the 5th, 6th and 7th are equal they would all be allocated the rank 6.

Even measures which cannot be measured numerically or exactly can be used in this context. All that is required is that the items can be put into an order of merit in respect of the quality being measured e.g. depth of colour or age of trees or rocks.

Example
Consider the distributions of reading speeds and intelligence quotients quoted in the earlier section of this chapter on scatter diagrams. The calculation of the coefficient of rank correlation can be tabulated as follows.

Child	Reading Speed	Intelligence Quotient	Rank in Reading Speed	Rank in Intelligence Quotient	d	d^2
A	120	80	11	12	-1	1
B	140	90	$8\frac{1}{2}$	$10\frac{1}{2}$	-2	4
C	100	90	12	$10\frac{1}{2}$	$+1\frac{1}{2}$	$2\frac{1}{4}$
D	170	100	7	$8\frac{1}{2}$	$-1\frac{1}{2}$	$2\frac{1}{4}$
E	130	100	10	$8\frac{1}{2}$	$+1\frac{1}{2}$	$2\frac{1}{4}$
F	190	105	5	7	-2	4
G	220	110	3	$5\frac{1}{2}$	$-2\frac{1}{2}$	$6\frac{1}{4}$
H	140	110	$8\frac{1}{2}$	$5\frac{1}{2}$	$+3$	9
I	180	115	6	4	$+2$	4
J	240	120	2	3	-1	1
K	200	125	4	2	$+2$	4
L	270	130	1	1	0	0

$$\sum d^2 = 40$$

$$\text{Coefficient of Rank Correlation} = R = 1 - \frac{6 \sum d^2}{n(n^2 - 1)}$$

$$= 1 - \frac{6 \times 40}{12 \times 143}$$

$$= 1 - 0.14$$

$$= \underline{0.86}$$

A high positive coefficient of correlation such as the above indicates that it is likely that the measures are closely linked but it must be remembered that no conclusive results can be obtained from such a small sample.

EXERCISE B

1. Calculate the coefficient of rank correlation for the following distributions of x and y (ref. covered example in section on drawing of regression lines)

x	10	15	23	30	35	45
y	11	19	18	27	29	33

2. Calculate the coefficient of rank correlation of the distributions listed in Question 3 of Exercise A of this Chapter.

3. Calculate the coefficient of rank correlation of the Distribution listed in Question 4 of Exercise A of this Chapter.

4. Calculate the coefficient of rank correlation of the distributions listed in Question 5 of Exercise A of this Chapter.

5. Calculate the coefficient of rank correlation of the distributions listed in Question 6 of Exercise A of this Chapter.

6. Calculate the coefficient of rank correlation of the distributions listed in Question 7 of Exercise A of this Chapter.

7. Six candidates sat a written examination and were also interviewed for the award of a bursary. The written examination was marked out of 100 and the interview was graded A, or B, or C, or D, or E in descending order of merit. The results were as follows:

Candidate	A	B	C	D	E	F
Written Examination	80	78	60	55	53	48
Interview Result	C	A	B	B	D	C

Find the coefficient of rank correlation for the two sets of measures.

8. The marks in Mathematics and French for a group of 10 children are listed below. Calculate the coefficient of rank correlation for these measures.

Pupil	A	B	C	D	E	F	G	H	I	J
Mathematics	18	32	44	47	56	71	76	84	86	93
French	40	68	53	35	84	75	58	26	48	77

9. The heights, weights and shoe sizes of a group of 15 senior boys and 15 senior girls, in a school, are shown below.

Girls Height (cm)	Weight (kg)	Shoe size
154	66	$5\frac{1}{2}$
170	57	$6\frac{1}{2}$
173	80	8
163	54	$4\frac{1}{2}$
160	53	5
164	54	5
163	53	6
160	62	$4\frac{1}{2}$
160	45	$4\frac{1}{2}$
163	53	5
168	60	7
162	55	$3\frac{1}{2}$
169	60	7
165	57	4
163	57	6

Boys Height (cm)	Weight (kg)	Shoe size
181	71	$9\frac{1}{2}$
173	70	8
184	69	$8\frac{1}{2}$
183	57	7
163	57	7
184	79	9
170	62	$8\frac{1}{2}$
180	73	9
165	54	$7\frac{1}{2}$
178	70	11
168	54	7
178	60	8
180	68	
180	65	8
173	70	$9\frac{1}{2}$

Calculate the coefficient of rank correlation for:
a) height against weight of girls
b) height against weight of boys
c) height against shoe size of girls
d) height against shoe size of boys.

10. Find the order of merit of the teams in the first division of the English Football League at the end of each of the last two years. (Ignore the teams which were relegated and promoted at the end of the first of these two years.) Find the coefficient of rank correlation for these two orders of merit. Repeat this exercise using the first division of the Scottish League.

PROBABILITY

What does probability mean to you?
The name rather suggests a vague likelihood
of something happening. There is nothing
vague however, about probability.
Mathematically, probability is as precise as
any other subject. We can calculate very
exactly the probability or chance of a certain
event happening when we know the
circumstances surrounding it.

If we toss a **fair** (or honest) coin it will
land head or tail upwards. There is a
'50–50' chance of either. In other words,
there is a probability of $\frac{1}{2}$ that a head shows
and the same for a tail.

If we are interested in obtaining a head,
then we say a head is a *favourable* outcome.
The total number of possible outcomes is 2
(a head or a tail). The probability of
obtaining a head is 1 out of 2 or $\frac{1}{2}$.

Let us consider rolling a die. The die could
show a 1, 2, 3, 4, 5, or 6, i.e. there are 6
possible outcomes when a die is thrown. If
we are interested in obtaining a 3, then there
is a chance of 1 in 6 or $\frac{1}{6}$ of getting a 3; there
is also a chance of $\frac{1}{6}$ of getting a 1 or 2 or 4
or 5 or 6.

The probability of a 3 showing
$$= \frac{\text{number of favourable outcomes}}{\text{number of possible outcomes}}$$
$$= \frac{1}{6}$$

What if we wanted to know the probability
of a number divisible by 2 showing when a
die is rolled? Again the total possible
number of outcomes is 6. Now we must

calculate the number of favourable outcomes
i.e. how many of the numbers are divisible by 2.
The numbers divisible by 2 are 2, 4 and 6.
Thus the probability of a number divisible
by 2 showing

$$= \frac{\text{number of favourable outcomes}}{\text{number of possible outcomes}}$$
$$= \frac{3}{6}$$
$$= \frac{1}{2}$$

So far, the probabilities we have considered
have been purely theoretical probabilities.
Let us now look at some actual results
obtained experimentally.

EXPERIMENT 1   Tossing a coin

You should all have a coin to toss and a book
and pencil to record your results.

Copy this table into your book.

Number of Tosses = 50

	Tally	Total
Heads		
Tails		

You must make sure before you begin that
you are tossing the coin *properly*. It must

spin round in the air before it falls. Now toss the coin 50 times and record the number of heads and tails obtained.

Tally your score as you did for frequency distributions. Finally total up your scores.

Let us compare the experimental results with what could be expected theoretically.

The probability of a head in one toss

$$= \tfrac{1}{2}$$

The expected number of heads in 50 tosses

$$= \tfrac{1}{2} \text{ of } 50$$
$$= 25$$

Similarly, the expected number of tails in 50 tosses is 25. Now, compare your experimental results with the theoretical results in this table.

Number of tosses = 50

	Experimental	Theoretical
Heads		25
Tails		25

Part 2
Repeat your experiment, tossing the coin 100 times. Again use a table to record the results. Calculate the theoretical results and compare them with the experimental results as in Part I.

Part 3
Let us now collect together the number of heads and tails for the whole class for the 100 tosses.

Copy this table into your book. Each member of the class should read out his results in turn and everyone should record all the results. As before, calculate the theoretical results and compare them in the table.

Total number of tosses =

	Experimental		Theoretical	
	Heads	Tails	Heads	Tails
Total				

It should have become obvious that the greater the number of tosses of a fair coin we consider, the nearer the experimental results come proportionally to the theoretical.

EXPERIMENT 2   Throwing a die

A similar experiment may be done using a die. Each member of the class should have a fair (honest) die and a shaker.

Part 1
Copy this table into your book.

Number of throws = 60

	Tally	Total
1		
2		
3		
4		
5		
6		

Roll the die 60 times and record as before the number of times 1, 2, 3, 4, 5, 6 turn up.

Now, compare the experimental and theoretical results in a Table.

The probability of a 1 in 1 throw

$$= \frac{\text{favourable}}{\text{possible}}$$

$$= \frac{1}{6}$$

The expected number of 1s in 60 throws

$$= \frac{1}{6} \text{ of } 60$$

$$= 10$$

Similarly, the expected number of 2s, 3s etc.

$$= 10$$

Number of throws = 60

	Experimental	Theoretical
1		10
2		10
3		10
4		10
5		10
6		10

Part 2
Repeat the experiment, rolling the die 120 times and recording your scores in a table as before. Calculate the theoretical results and compare them, in a table, with the experimental results.

Part 3
As in Experiment 1, collect together the total numbers of 1s, 2s, 3s etc. that the whole class had for Part 2. Again calculate the expected theoretical values. Complete this table.

Total Number of Throws =

1	2	3	4	5	6

Experimental totals

Theoretical totals

It should be clear, again, that the experimental results come closer proportionally to the theoretical results as we use a larger number of throws of the die.

Further experiments are quite easily devised for comparing experimental and theoretical results.

Suggestions
1. A solid regular triangular pyramid (tetrahedron) with the numbers 1, 2, 3, 4 painted on its faces may be rolled in a similar way to a die.
2. A spinner, as shown, can be made easily (but care must be taken). The disc may be marked off into any number of equal segments and either coloured or numbered as wished. The arrow should be able to spin round freely.

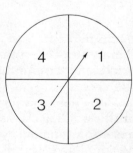

With a spinner of this type, a great many different experiments may be devised.
3. Various experiments may be conducted using books. e.g. opening a book at random and counting the number of times the tens digit is odd (or even) or a multiple of 2 or 3 etc.
4. On the market is a small roulette wheel that people can use at home. If some member of the class could borrow one of these, many interesting probability experiments could be tried.

Bias

You may have noticed that in the experiments with the coin and the die, the word 'honest' or 'fair' was used.

An honest coin is one which is equally likely to show a head or a tail.

An honest die is one in which each number has an equal chance of turning up.

A coin or die which is not honest is said to be **biased** and it will give biased results.

A two-headed coin would obviously be biased since there is no chance at all of getting a tail.

Dice that are used for gambling can be 'loaded' or weighted so that they tend to land a certain way up.

In a later chapter you will learn how you can test the honesty of a coin or a die, if there are any doubts about it.

In the suggested experiments, you must be wary of introducing any bias. The spinner, for example, would have to be made very carefully. The material of the platform must be of uniform thickness and the circle perfectly round. The segments also must be exactly equal. Another place where great care is required, is in the spinning arrow. It must be able to spin round quite freely.

In the experiment with the book, the book you choose to use, should not be one that has been habitually opened at certain pages or you will find that it opens at these pages more or less automatically. So, examine the book first to ensure that this sort of thing is not likely to happen.

Calculating probabilities

One of the difficulties in calculating probabilities is in working out the number of favourable outcomes and the number of possible outcomes. Each example must be considered carefully. Once we know these two quantities, we can very easily calculate the probability using this formula.

Probability of an event happening

$$= \frac{\text{Number of favourable outcomes}}{\text{Number of possible outcomes}}$$

If an event is certain to occur then the number of favourable outcomes is equal to the number of possible outcomes thus:
The probability of an event which is certain to occur = 1.

If an event is certain not to happen then the number of favourable outcomes is 0 and thus:
The probability of an event which is certain not to occur = 0.

If the probability that an event occurs is p and the probability that the event does not occur is q then:

$$p + q = 1 \text{ or } q = 1 - p.$$

Example 1
A bag contains 50 marbles, 40 black ones and 10 red.
  a) What is the probability of drawing a red marble?
  b) What is the probability of drawing a black marble?
  c) If two black marbles are removed from the bag what is the chance now of drawing a black one?
  d) What is the chance of drawing a white marble?
  e) What is the probability of drawing either a red marble or a black one?

Solution
a) Number of favourable
   outcomes   = 10
   Number of possible
   outcomes   = 50

Probability of drawing a
red marble

$= \dfrac{\text{favourable}}{\text{possible}}$

$= \dfrac{10}{50}$

$= \dfrac{1}{5}$

b) Number of favourable
outcomes $= 40$

Number of possible
outcomes $= 50$

Probability of drawing a
black marble

$= \dfrac{\text{favourable}}{\text{possible}}$

$= \dfrac{40}{50}$

$= \dfrac{4}{5}$

c) Number of favourable
outcomes $= 38$

Number of possible
outcomes $= 48$

Probability of drawing a
black marble

$= \dfrac{38}{48}$

$= \dfrac{19}{24}$

d) Number of favourable
outcomes $= 0$

Number of possible
outcomes $= 50$

Probability of drawing a
white marble

$= \dfrac{\text{favourable}}{\text{possible}}$

$= \dfrac{0}{50}$

$= 0$

(i.e. there is no possible
chance of drawing a white
marble)

e) Number of favourable
outcomes $= 50$

Number of possible
outcomes $= 50$

Probability of drawing a
red or black marble

$= \dfrac{\text{favourable}}{\text{possible}}$

$= \dfrac{50}{50}$

$= 1$

(i.e. it is a certainty that
either a red or a black
marble is drawn).

Example 2

Two dice thrown at the same time. What is
the probability that the sum of the two
numbers shown is six?

Solution

In this example, the difficulty lies in working
out all the possible outcomes. The simplest
way to do this, is to write down, in order, all
the pairs of numbers which could possibly
show on the dice. The first number refers to
the first die and the second to the other die.

1, 1	2, 1	3, 1	4, 1	5, 1	6, 1
1, 2	2, 2	3, 2	4, 2	5, 2	6, 2
1, 3	2, 3	3, 3	4, 3	5, 3	6, 3
1, 4	2, 4	3, 4	4, 4	5, 4	6, 4
1, 5	2, 5	3, 5	4, 5	5, 5	6, 5
1, 6	2, 6	3, 6	4, 6	5, 6	6, 6

Thus we see that there are 36 possible
outcomes, and 5 favourable outcomes.

Probability that the sum of
the numbers is 6

$= \dfrac{\text{favourable}}{\text{possible}}$

$= \dfrac{5}{36}$

EXERCISE A

1. A bag contains 30 marbles, 25 green ones and 5 red. What is the chance of picking out a green marble?

2. If a letter is taken at random from the word CHRYSANTHEMUM, what is the probability that
    a) it is a vowel?
    b) it is 'M'?

3. On throwing a die, what is the probability of turning up
    a) a 6?
    b) a number less than 3?
    c) a number more than 6?
    d) an odd number?

4. If we take a standard pack of 52 playing cards, what is the chance of drawing
    a) an ace?
    b) the ace of clubs?
    c) a heart?
    d) a king, queen or jack?
    e) a joker?

5. A black card is removed from a pack of cards. What is the probability of drawing
    a) a black card?
    b) a red queen?
    c) a king?

6. If a number is chosen at random from the numbers 1 to 30 inclusive, what is the chance that a prime number is picked?

7. A box of 2 dozen pencils contains 6 with broken points. What is the probability of picking out one which has not got a broken point?

8. In a car park, there are 100 vehicles, 85 of them being cars, 10 lorries and 5 buses. If they are all equally likely to leave, what is the probability of
    a) a bus leaving first?
    b) a car leaving second? (If a bus left first).

9. A box contains 50 coloured pencils, 20 red, 15 blue and the rest green. What is the chance of drawing
    a) a red pencil?
    b) a blue or green pencil?

    c) a yellow pencil?
    d) a red, blue or green pencil?

10. Two dice are thrown. What are the probabilities that the total score is
    a) 5?
    b) 1?
    c) 10?
    d) 14?
    e) less than 13?

11. Two pennies are tossed together. What is the probability of two heads showing?

12. If three pennies are tossed together, what is the chance of
    a) three heads showing?
    b) less than 2 tails showing?
    c) more than 2 tails showing?

13. A coin and a die are thrown together. What is the probability of
    a) a head and a 3 showing?
    b) a tail and a 4 showing?
    c) a tail and a 7 showing?

14. A solid regular triangular pyramid (tetrahedron) has the numbers 1, 2, 3 and 4 marked on its 4 faces. If it is thrown like a die, what is the probability that
    a) the 1– face lands downwards?
    b) the sum of the three faces showing is an odd number?

15. If 2 tetrahedrons like the one in the last question are thrown together, what is the chance that
    a) the sum of the numbers face down is odd?
    b) the sum of the numbers face down is a prime number?

16. If a tetrahedron and a coin are thrown together, what is the probability of getting
    a) an odd number and a head showing?
    b) a multiple of 2 and a tail?

17. A die and 2 coins are tossed together. What is the chance of obtaining
    a) 2 heads and an even number?
    b) 1 head, 1 tail and a number divisible by 3?
    c) 2 tails and a number less than 3?
    d) 1 head, 1 tail and a number greater than 3?

18. A pair of dice and a penny are thrown together. Find the probability of
   a) a head and a total score of 7 showing.
   b) a tail and a total score greater than 10 showing.
   c) a tail and a total score less than 13 showing.

19. Four coins are tossed together. What is the probability of obtaining 4 heads?

20. Considering question 12 and question 19 could you calculate the probability of getting 5 heads when 5 coins are tossed together, without making an array of the results?

Multiplication law

When concerned with mutually exclusive events it is not always necessary to write out an array e.g. consider the experiment of tossing a penny and throwing a die to find the probability of getting a head and a two. In half of the trials we should expect to get a head and within that half, the distribution of scores on the die is expected to be uniform and thus we expect to get a favourable result in $\frac{1}{6}$ of that half of the trials.

Therefore the probability of getting a head and a two is $\frac{1}{2} \times \frac{1}{6} = \frac{1}{12}$.

Generalising this concept we have that if two mutually exclusive events E_1 and E_2 have probabilities of occurring p_1 and p_2 then the probability that both events will occur is $p_1 \times p_2$.

Example 1
What is the probability of obtaining 3 heads in 3 tosses of a coin?

Solution
Probability of a head on 1st toss $= \frac{1}{2}$

Probability of a head on 2nd toss $= \frac{1}{2}$

Probability of a head on 3rd toss $= \frac{1}{2}$

$\therefore$ Probability of 3 heads $= \frac{1}{2} \times \frac{1}{2} \times \frac{1}{2}$

$= \frac{1}{8}$

Example 2
What is the chance of throwing 2 sixes with two throws of a die?

Solution
Probability of a 6 on 1st throw $= \frac{1}{6}$

Probability of a 6 on 2nd throw $= \frac{1}{6}$

$\therefore$ Probability of 2 sixes $= \frac{1}{6} \times \frac{1}{6}$

$= \frac{1}{36}$

You may check that the multiplication law gives the correct answers by writing out all the possible outcomes.

In *Example 1*, the possibilities are as follows:

1st toss	2nd toss	3rd toss
H	H	H
H	H	T
H	T	H
H	T	T
T	H	H
T	H	T
T	T	H
T	T	T

In all there are 8 possibilities

Probability of 3 heads in three tosses

$= \dfrac{\text{favourable}}{\text{possible}}$

$= \dfrac{1}{8}$

84

In *Example 2* the possibilities are as follows:

1, 1	2, 1	3, 1	4, 1	5, 1	6, 1
1, 2	2, 2	3, 2	4, 2	5, 2	6, 2
1, 3	2, 3	3, 3	4, 3	5, 3	6, 3
1, 4	2, 4	3, 4	4, 4	5, 4	6, 4
1, 5	2, 5	3, 5	4, 5	5, 5	6, 5
1, 6	2, 6	3, 6	4, 6	5, 6	6, 6

The probability of 2 sixes $= \dfrac{\text{favourable}}{\text{possible}}$

$$= \frac{1}{36}$$

Example 3

Suppose we have a bag containing 10 balls, 7 red ones and 3 black.

a) What is the probability of picking out a red ball followed by a black one?

b) What is the probability that if we choose 2 balls, one will be red and one will be black?

Solution

a) Probability of picking a red ball $= \dfrac{\text{favourable}}{\text{possible}}$

$$= \frac{7}{10}$$

Now, there are only 9 balls left 6 red and 3 black

Probability of picking a black ball $= \dfrac{\text{favourable}}{\text{possible}}$

$$= \frac{3}{9}$$

$$= \frac{1}{3}$$

Probability of picking a red ball followed by a black one $= \dfrac{7}{10} \times \dfrac{1}{3}$

$$= \frac{7}{30}$$

b) This must be taken in 2 parts— assuming that a red ball is picked first and then assuming that a black ball is picked first. If we pick a red ball first, followed by a black one, the probability is $\dfrac{7}{30}$ (See *a*) above).

BUT, we could have picked a black ball first.
Probability of picking a black ball $= \dfrac{\text{favourable}}{\text{possible}}$

$$= \frac{3}{10}$$

Now there are 9 balls left, 7 red and 2 black
Probability of picking a red ball $= \dfrac{\text{favourable}}{\text{possible}}$

$$= \frac{7}{9}$$

Probability of picking a black ball followed by a red ball $= \dfrac{3}{10} \times \dfrac{7}{9}$

$$= \frac{7}{30}$$

Hence, the total probability of picking a red ball and a black ball (in any order)

$$= \frac{7}{30} + \frac{7}{30}$$

$$= \frac{7}{15}$$

EXERCISE B

1. What is the probability of throwing 6 tails in 6 tosses of a coin?
2. What is the probability of obtaining a total of two with two throws of a die?
3. Out of 10 steel valve springs, 3 are defective. Two springs are chosen at random for testing. What is the probability that both test springs are
   a) not defective?
   b) defective?

4. In question 3, what is the chance of not getting a defective spring if 3 springs are tested?

5. What is the probability of drawing a red queen followed by a black king from a standard pack of cards?

6. In a box, there are 25 coloured pencils; 10 are red, 8 are blue and 7 green. What is the probability of picking
   a) a red pencil followed by a blue one?
   b) a red and a green pencil?
   c) 2 blues and a green?

7. There are 50 marbles in a bag, 20 white ones, 20 blue ones, 5 clear ones and 5 yellow ones. What is the probability of picking out
   a) 1 of each colour?
   b) 2 blue ones and a white one?
   c) 2 clear marbles and 2 yellow ones?
   d) 4 white marbles?

Calculating expected frequencies

The idea of calculating expected frequencies was introduced earlier in the chapter, in the experimental work.

If the probability of a head turning up when a coin is tossed is $\frac{1}{2}$, then when a coin is tossed 10 times you could expect 5 heads to show.

If the probability of obtaining a 6 when a die is thrown is $\frac{1}{6}$, then in 60 throws of a die you could expect $\frac{1}{6}$ of 60 or 10 sixes to show.

The expected frequency = probability of the event happening × number of trials.

Example

A die is thrown 50 times. What is the expected frequency of an even number turning up?

$$\text{Probability of an even number} = \frac{\text{favourable}}{\text{possible}}$$

$$= \frac{3}{6}$$

$$= \frac{1}{2}$$

The expected frequency of even numbers $= \frac{1}{2}$ of 50

$$= 25$$

1. If a coin is tossed 100 times, how many tails might you expect to turn up?

2. When a die is thrown 60 times, what is the expected frequency of
   a) a number less than 4 showing?
   b) a number less than 7 showing?
   c) a factor of 6 showing?

3. If a tetrahedron, with its faces marked 1 to 4, is thrown 80 times, how many times would you expect
   a) the two-face to be down?
   b) the sum of the faces showing to be more than 7?

4. Two tetrahedrons are thrown 160 times. How many times would you expect the sum of the 2 faces that land downward to be an odd number?

5. What is the expected frequency of two heads showing, if 2 coins are tossed together 80 times?

6. A pack of cards is shuffled and then a card is drawn. This card is replaced and the experiment is repeated another 149 times. Approximately how many times would you expect to pick
   a) a face card?
   b) a black card?

7. A town has 1250 children under the age of 10 years who have never had chickenpox. If it is known, from experience, that the probability of a child under 10 catching chickenpox is 0.16, how many of these children would you expect to take chickenpox?

8. In a certain town, 10% of the car drivers have some kind of accident, however slight, in a year's driving. If there are 3672 drivers in the town, what is the expected number of accidents in the coming year?

9. If the probability of having defective hearing is 0.08 in school children, how many pupils in a school of 1000 could be expected to have defective hearing?

10. Imagine the chance of a footballer breaking a leg during a football season is 0.15. If 60 players sign on with a club at the start of the season, how many of them are likely to break a leg during the season?

11. If the probability that a child has taken measles by the age of 12, is 0.6, how many of the 300 children aged 12 in a school could be expected not to have had measles?

9 PERMUTATIONS AND COMBINATIONS

In order to reduce the labour of tabulating in detail all the ways in which a sequence of events can occur we turn to the study of permutations and combinations.

If an event A can occur in n_1 ways and if for each of them an event B can thereafter occur in n_2 ways then events A and B can together occur in that order in $n_1 \times n_2$ ways, e.g. if there are 3 routes from Newburgh to Newton and 4 routes from Newton to Oldtown then there are 3×4 routes from Newburgh to Oldtown via Newton.

This result can be generalised and thus if there are r events which can take place in order and if the first event can occur in n_1 ways and thereafter the second event can occur in n_2 ways and so on until the r th event which can occur in n_r ways then the r events can occur in order in $n_1 \times n_2 \times n_3 \times \ldots \times n_r$ ways.

Permutations

If we have r spaces and n objects, all different, with which to fill the spaces and proceed to form all the different ways in which this can be done it is said that we have formed all the permutations of the n objects taken r at a time and the number of ways that this can be done is called the number of permutations of n objects taken r at a time and is written ^nP_r. For example if we have a committee of four members and wish to select a convenor and secretary from the committee, we can tabulate all the possible ways in which this can be done as shown at the bottom of the page.

The number of ways in which the selection can be made $= {}^4P_2 = 12$. It can also be noted that any of the four members of the committee could be chosen as the convenor and thereafter for each of the four possible choices there are three members from whom we can choose a secretary. Thus there are four choices of convenor and for each of these there are three choices of secretary i.e.

Number of permutations $= {}^4P_2 = 4 \times 3$
$$= /12.$$

Value of ^nP_r

In general if we have n objects and wish to fill r spaces then there are n choices of object with which to fill the first space and for each of these choices there are $(n-1)$ remaining objects from which we can make our choice of object to fill the second space and thus there are $n(n-1)$ ways of filling the first two spaces. For each of these $n(n-1)$

Convenor	A	A	A	B	B	B	C	C	C	D	D	D
Secretary	B	C	D	A	C	D	A	B	D	A	B	C

ways of filling the first two spaces there are $(n-2)$ remaining objects from which we can make our choice of object to fill the third space and thus there are $n(n-1)(n-2)$ ways of filling the first three spaces. This method can be continued until we come to the rth space when there are $(n-r+1)$ objects left from which we can make our choice and thus there are

$n(n-1)(n-2) \ldots \ldots (n-r+1)$ ways of filling the r spaces.

i.e.

$$^nP_r = n(n-1)(n-2) \times \ldots \ldots \times (n-r+1)$$

$$= \frac{n!}{(n-r)!}$$

(Where $n! = n(n-1)(n-2) \times \ldots \ldots \times 2 \times 1$.)

Note: i) The number of ways in which n objects can be permuted among themselves

$$= {}^nP_n = n! = n(n-1)(n-2) \times \ldots \ldots \times 2 \times 1.$$

ii) If our problem places restrictions upon the use of some of the objects it is usually good policy to deal with the most restricted objects first.

Permutations when the objects are not all different

Consider n objects of which r are alike of one kind and all the others are different and let us suppose that there are x permutations of these n objects among themselves. Take one of these permutations and let us make the r objects which were alike distinguishable and they can now, while holding the same set of positions in the arrangement, be permuted among themselves in r! ways. If this process is repeated for all of the x arrangements then we will obtain all of the permutations of n objects among themselves thus

$$x \times r! = n!$$

i.e. $$x = \frac{n!}{r!}$$

If the above concept is extended to the case of n objects of which p are alike of one kind,

q are alike of a second kind and r are alike of a third kind and all of the other objects are different then the number of permutations of the n objects among themselves is given by the formula

$$\frac{n!}{p! \times q! \times r!}$$

Example 1
Find the number of permutations of the letters in the word HEADS among themselves and find the probability that an arrangement chosen at random will start and end with a vowel.

Solution
Number of permutations of the letters
$$= 5! = 5 \times 4 \times 3 \times 2 = 120.$$
Number of permutations which start and end with a vowel
$$= 2 \times 1 \times 3!$$
$$= 12.$$

(There are two choices of a vowel to fill the first place and for each of these choices there is one choice of a vowel for the last place and thereafter the other three letters can be permuted among themselves in 3! ways.)

$$\text{Required Probability} = \frac{12}{120} = \frac{1}{10}$$

Example 2
Find the number of ways there are of selecting the 1st, 2nd and 3rd prize winners from a beauty contest in which there are 15 contestants.

Solution
Number of permutations $= {}^{15}P_3$

$$= \frac{15!}{(15-3)!}$$

$$= \frac{15!}{12!}$$

$$= 15 \times 14 \times 13$$

$$= 2730.$$

Example 3
Find the number of ways of permuting the letters of the word LETTERED among themselves.

Solution
Number of permutations

$$= \frac{8!}{2! \times 3!}$$

$$= \frac{8 \times 7 \times 6 \times 5 \times 4 \times 3 \times 2 \times 1}{2 \times 1 \times 3 \times 2 \times 1}$$

$$= 8 \times 7 \times 6 \times 5 \times 2$$

$$= 3360$$

Exercise A

1. Evaluate $^5P_2$, $^9P_3$, $^7P_7$, $^8P_5$, $^nP_2$.

2. On a lunch menu there are three choices of first course, five choices of second course and four choices of third course. Find the number of different three course meals that can be chosen.

3. Find how many different 4 digit numbers can be formed using one of each of the digits 1, 2, 3, 4. If one of these numbers is selected at random find the probability that it will be less than 2000.

4. Find the number of permutations of the letters of the word PERMUTE among themselves.

5. Find the number of permutations of 4 ps and 3 qs among themselves. If one of these arrangements is chosen at random find the probability that the 3 qs will come together.

6. Find the number of ways of selecting a captain and vice-captain from a team of 11 players.

7. In a competition the entrant is given a list of 10 kitchen gadgets and has to select six of them and place them in order of importance. Find the number of different selections that could be made and state the probability that any one selection is correct.

8. 4 passengers enter a railway carriage which seats six people. If one of them insists on facing the engine find the number of ways in which they can select their seats.

9. I have 10 shirts in my wardrobe and plan to wear a different one each day of the week. In how many different ways can I make my selection of shirts for the various days of the week?

10. Any football match must end in a home win, an away win or a draw. How many different outcomes are there for a list on 10 matches?
What is the probability that my forecast of the list of results is correct?

11. There are 9 empty seats in the classroom and 6 new pupils join the class. In how many different ways could they choose to take their seats?

12. Using the digits 1, 1, 2, 3, 4, not more than once each, find how many numbers greater than 2000 can be formed.

13. Find the probability that if you choose a 4 digit number it will contain,
1) one of each of the digits 0, 1, 2, 3,
2) only the digits 0, 1, 2, 3, repetitions of a digit being included.

Combinations

In the last section we were always interested in the order in which a selection was made but frequently the order is of no importance e.g. the selection of a committee from a club or the selection of a hand of cards from a pack. In these cases we have no interest in the order of selection but only in the final result of the selection and here we say that the number of ways in which this can be done is the number of combinations of, say, n objects taken r at a time and we denote it by the symbol ^nC_r.

Value of ^nC_r

Since the only thing that is of importance in a combination of items is the actual set of items in the selection all of the selections of n objects taken r at a time will contain a different set of r items. If we now take each of these selections of r items and permute these r items among themselves, this can be done in r! ways, and will give all the permutations of n items taken r at a time thus

$$^nC_r \times r! = {}^nP_r$$

i.e.
$$^nC_r = \frac{^nP_r}{r!}$$

$$= \frac{n!}{(n-r)! \times r!}$$

Example 1

Find the number of ways of selecting a set of three books from a shelf containing eight different books.

Solution

Number of ways of selecting the books =
$$^8C_3 = \frac{8!}{(8-3)! \times 3!}$$

$$= \frac{8!}{5! \times 3!}$$

$$= \frac{8 \times 7 \times 6 \times 5 \times 4 \times 3 \times 2 \times 1}{5 \times 4 \times 3 \times 2 \times 1 \times 3 \times 2 \times 1}$$

$$= 8 \times 7 = \underline{56}$$

Example 2

Find the number of ways of selecting a committee of 3 ladies and 2 gents from a club containing 7 ladies and 5 gents.

Solution

Number of ways of selecting the ladies =
$$^7C_3 = \frac{7!}{(7-3)! \times 3!}$$

$$= \frac{7 \times 6 \times 5 \times 4 \times 3 \times 2 \times 1}{4 \times 3 \times 2 \times 1 \times 3 \times 2 \times 1}$$

$$= 7 \times 5 = \underline{35}.$$

Number of ways of selecting the gents =
$$^5C_2 = \frac{5!}{(5-2)! \times 2!}$$

$$= \frac{5 \times 4 \times 3 \times 2 \times 1}{3 \times 2 \times 1 \times 2 \times 1}$$

$$= 5 \times 2 = \underline{10}.$$

Total number of ways of selecting the committee
$$= 35 \times 10 = \underline{350}.$$

EXERCISE B

1. Evaluate $^5C_3$, $^7C_4$, $^8C_2$, $^6C_5$, $^8C_6$, $^6C_1$.

2. Find the number of ways of selecting a committee of 3 people from a group of 9 people.

3. Find the number of ways in which you can select a hand of 4 cards from a pack of 52 cards.

4. Find the number of ways in which you can select, from the school library, 3 Dickens' novels for Christmas reading if there are 10 novels by Dickens in the library.

5. Find the number of ways in which the teacher can select 3 examples, for use as a test, from an Exercise of 12 examples.

6. Find the number of ways in which you can select a committee of 5 from a group of 8 men and 7 women if there must not be less than two members of either sex on the committee.

7. Find the number of ways in which you can split a group of 7 boys into two groups, one of 4 and the other of 3 boys.

8. As part of a survey I am required to visit 20% of the houses in a particular street. If there are 20 houses in the street in how many different ways can I make my selection?

9. In how many different ways can I select 6 rose trees, all different, from a catalogue list containing 12 varieties?

10. In question 2 find the number of ways in which the committee could be selected if two particular people refuse to serve on the same committee. If a committee from question 2 were selected at random find the probability that these two people would have been asked to serve on the same committee.

11. A delivery boy is given 9 items, all different, and is required to deliver 4 of them at house A, 3 at house B and two at house C. If he delivers the items at random find the number of ways in which he can do the job and the probability that he does it wrongly.

12. Find the number of ways in which you can select two items from a set of eight items. If a blind man selects two socks from a drawer containing four pairs of socks find the probability that he selects a matching pair.

10 THE BINOMIAL DISTRIBUTION

The word binomial contains the prefix *bi* meaning two. In the Binomial distribution we are concerned with *two* possible events occurring. If a coin is tossed, it will either come down heads or tails. If you go out in the street, the next person you meet will be either male or female. Next Tuesday will either be your birthday or not. In each case there are two possibilities.

If we decide which of the two possibilities is desirable, then there is a probability of its happening and a probability of its not happening. In Statistics, we normally say that the probability of a certain event happening is **p** and of its not happening is **q**.

Suppose we have a bag containing 100 balls, some red and some white. If there are 50 of each colour, we have an equal chance of drawing a white one or a red one. Thus the probability of obtaining a red ball is 50% or $\frac{1}{2}$ or 0.5, and the probability of obtaining a white one is the same. The two probabilities add up to 1.

Let us now suppose that in the bag there are 90 red balls and 10 white ones. The probability of drawing a red one is now $\frac{90}{100}$ or $\frac{9}{10}$ or 0.9, and the probability of failing to draw a red one is $\frac{10}{100}$ or $\frac{1}{10}$ or 0.1. Again the two probabilities add up to 1.

If p is the probability of drawing a red ball and q is the probability of not drawing a red ball, then in each of the above cases, we can say

$$p + q = 1$$

Consider another example. A school caretaker was clearing out old text books and he put a large number of English books and Geography books into the same sack, 30% of the books being English books and the rest Geography books.

Let p = probability of picking out an English book.

Let q = probability of picking out a Geography book.

In this case, $p = \frac{30}{100}$ or $\frac{3}{10}$ or 0.3

and, $q = \frac{70}{100}$ or $\frac{7}{10}$ or 0.7

Again $p + q = 1$ if we select one book. Suppose we want to pick out two books. This time there are four possible outcomes: 2 English books, 1 English book followed by 1 Geography book, 1 Geography book followed by 1 English book or 2 Geography books. The probabilities of the various possible outcomes are thus given by the formulae $p \times p$, $p \times q$, $q \times p$, $q \times q$ and we can further note that these terms can be further grouped as p^2, $2pq$, q^2 where p^2 is the probability of getting two English books, $2pq$ is the probability of getting one of each kind and q^2 is the probability of getting two Geography books. Since these are the only possible outcomes of our experiment their total must be 1 and thus we note that the terms of the formula
$$(p + q)^2 = p^2 + 2pq + q^2 = 1$$
give the probabilities of the various outcomes of carrying out the experiment twice.

If we extend the above to the selection in turn of three books we have the position whereby each of the above results can be followed by the selection of an English or a Geography book. Thus the probability of selecting three English books is $p \times p \times p = p^3$. Secondly there are three ways in which we can select two English and one Geography books: the Geography one can be selected either first, second or third, and the probabilities of these various outcomes are $q \times p \times p$, $p \times q \times p$, $p \times p \times q$ and the total of these three terms, namely $3p^2q$ gives the probability of selecting two English and one Geography books. Similarly we can show that the probability of selecting one English and two Geography books is $3pq^2$ and of selecting three Geography books is q^3. The probabilities of the various possible outcomes of three trials are given by the expansion

$$(p+q)^3 = p^3 + 3p^2q + 3pq^2 + q^3 = 1.$$

This concept can be extended and thus the probabilities of the various outcomes of n trials of an experiment are given by the terms of the expansion of $(p+q)^n$.

Expansion of $(p+q)^n$

When dealing with low values of n it is reasonable to do the actual multiplications from first principles but for higher values it is essential that we look for a better technique.

If we consider the first few of the expansions of $(p+q)^n$ we obtain the following expansions:

$$(p+q)^1 = p+q$$
$$(p+q)^2 = p^2 + 2pq + q^2$$
$$(p+q)^3 = p^3 + 3p^2q + 3pq^2 + q^3$$
$$(p+q)^4 = p^4 + 4p^3q + 6p^2q^2 + 4pq^3 + q^4$$

The experience of the last chapter may lead us to see that the coefficients of the terms in these expansions can be written in terms of the ^nC_r symbols, e.g.

$$(p+q)^4 = p^4 + {}^4C_1p^3q + {}^4C_2p^2q^2 + {}^4C_3pq^3 + {}^4C_4q^4,$$

and this can lead us to the assumption that the general expansion is:

$$(p+q)^n = p^n + {}^nC_1p^{n-1}q + {}^nC_2p^{n-2}q^2 + \ldots + {}^nC_rp^{n-r}q^r + \ldots {}^nC_nq^n.$$

We shall now proceed to prove the truth of this expansion by means of induction. Let us assume that the formula is true for some value of n, say $n = m$, i.e. we assume that

$$(p+q)^m = p^m + {}^mC_1p^{m-1}q + {}^mC_2p^{m-2}q^2 + \ldots + {}^mC_rp^{m-r}q^r + \ldots {}^mC_mq^m.$$

then

$$(p+q)^{m+1} = (p+q)^m(p+q)$$
$$= (p^m + {}^mC_1p^{m-1}q + {}^mC_2p^{m-2}q^2 + \ldots + {}^mC_{r-1}p^{m-r+1}q^{r-1} + {}^mC_rp^{m-r}q^r + \ldots + {}^mC_mq^m)(p+q)$$

It is impossible for us to carry out this multiplication for the general case but we can consider individual terms of the product. Let us take as our general term the one which involves q^r. We shall obtain two terms in the product which involve q^r and these will be obtained from the product of $^mC_{r-1}p^{m-r+1}q^{r-1}$ and q and the product of $^mC_rp^{m-r}q^r$ and p. Thus the general term in the product is

$${}^mC_{r-1}p^{m-r+1}q^r + {}^mC_rp^{m-r+1}q^r$$
$$= ({}^mC_{r-1} + {}^mC_r)p^{m-r+1}q^r$$
$$= {}^{m+1}C_rp^{m+1-r}q^r$$

(Note that it can be proved that $^mC_{r-1} + {}^mC_r = {}^{m+1}C_r$)

Applying this result to all of the terms of our products we obtain

$$(p+q)^{m+1} = p^{m+1} + {}^{m+1}C_1p^mq + {}^{m+1}C_2p^{m-1}q^2 + \ldots + {}^{m+1}C_rp^{m+1-r}q^r + \ldots + {}^{m+1}C_{m+1}q^{m+1}.$$

i.e. if the expansion is true for $n = m$ then it is also true for $n = m+1$, but the expansion was true for $n = 1$ and by the above it must thus be true for $n = 1+1 = 2$. Since it is true for $n = 2$ by the above it must also be true for $n = 2+1 = 3$ and by continuing this argument we can obviously prove the rule true for all positive integer values of n.

Thus for all positive integer values of n we have that:

$$(p+q)^n = p^n + {}^nC_1 p^{n-1}q + {}^nC_2 p^{n-2}q^2$$
$$+ \ldots + {}^nC_r p^{n-r}q^r + \ldots + {}^nC_n q^n.$$

and in this expansion the term ${}^nC_r p^{n-r}q^r$ gives the probability of getting $n-r$ successes and r failures in n trials of an experiment.

Pascal's triangle

Since some students may find the algebra of the previous section to be beyond the level of their interest we can now look at the number pattern known as Pascal's Triangle and use it to obtain the coefficients of $(p+q)^n$. This pattern of numbers is built up and can be proved a valid method of obtaining the coefficients by the same methods as we used in the last section.

It will be seen that each term in the triangle is found by adding together the two terms in the line above which lie on either side of it.

Pascal's Triangle, as stated gives the coefficients of the terms in the expansion of $(p+q)^n$

$$(p+q)^1 = 1p + 1q$$
$$(p+q)^2 = 1p^2 + 2pq + 1q^2$$
$$(p+q)^3 = 1p^3 + 3p^2q + 3pq^2 + 1q^3$$
$$(p+q)^4 = 1p^4 + 4p^3q + 6p^2q^2 + 4pq^3 + 1q^4$$
etc.   etc.

The way in which Pascal's Triangle is used is shown in the following examples.

Example 1
In a certain collection of bulbs 80% are daffodils and the rest are narcissi. Calculate the probability of obtaining 0, 1, 2, 3, and 4 narcissi in a group of 4 bulbs.

Solution

$$p = \frac{4}{5} = \text{daffodils}$$

$$q = \frac{1}{5} = \text{narcissi}$$

	PASCAL'S TRIANGLE	
Number of Trials (n)	Coefficients in the expansion of $(p+q)^n$	Number of Different Combinations
1	1 1	2
2	1 2 1	3
3	1 3 3 1	4
4	1 4 6 4 1	5
5	1 5 10 10 5 1	6
6	1 6 15 20 15 6 1	7
7	1 7 21 35 35 21 7 1	8
8	etc.	9
etc.		etc.

It may already have become obvious that the number of different combinations is always 1 greater than the number of trials of the experiment. E.g. if we wish to pick out 6 books together, there are 7 possible different combinations of English and Geography books. (Work out the combinations and check for yourself.)

With 4 bulbs, the expansion is $(p+q)^4 = 1$
The coefficients from Pascal's Triangle are 1, 4, 6, 4, 1.

Thus,
$$(p+q)^4 = p^4 + 4p^3q + 6p^2q^2 + 4pq^3 + q^4$$

By substituting $p = \frac{4}{5}$ and $q = \frac{1}{5}$, we get

95

a) Probability of 0 narcissus = p^4 = $\frac{4}{5} \times \frac{4}{5} \times \frac{4}{5} \times \frac{4}{5}$ = $\frac{256}{625}$

b)   ,,   ,, 1 ,, = $4p^3p$ = $\frac{4}{1} \times \frac{4}{5} \times \frac{4}{5} \times \frac{4}{5} \times \frac{1}{5}$ = $\frac{256}{625}$

c)   ,,   ,, 2 ,, = $6p^2q^2$ = $\frac{6}{1} \times \frac{4}{5} \times \frac{4}{5} \times \frac{1}{5} \times \frac{1}{5}$ = $\frac{96}{625}$

d)   ,,   ,, 3 ,, = $4pq^3$ = $\frac{4}{1} \times \frac{4}{5} \times \frac{1}{5} \times \frac{1}{5} \times \frac{1}{5}$ = $\frac{16}{625}$

e)   ,,   ,, 4 ,, = q^4 = $\frac{1}{5} \times \frac{1}{5} \times \frac{1}{5} \times \frac{1}{5}$ = $\frac{1}{625}$

(Check: $(p+q)^4$ must equal 1

$$\frac{256}{625} + \frac{256}{625} + \frac{96}{625} + \frac{16}{625} + \frac{1}{625} = \frac{625}{625} = 1$$

Example 2

If 6 pennies are tossed together, what is the probability of 6 tails occurring?

Solution

$$p = \frac{1}{2} = \text{heads}$$

$$q = \frac{1}{2} = \text{tails}$$

With 6 pennies, the expansion is
$$(p+q)^6 = 1$$

From Pascal's Triangle, the relevant coefficients are 1, 6, 15, 20, 15, 6, 1.

Thus
$(p+q)^6 = p^6 + 6p^5q + 15p^4q^2$
$\qquad\qquad + 20p^3q^3 + 15p^2q^4 + 6pq^5 + q^6$

q^6 represents 6 tails.

Probability of 6 tails = $q^6 = (\frac{1}{2})^6$

$$= \frac{1}{2} \times \frac{1}{2} \times \frac{1}{2} \times \frac{1}{2} \times \frac{1}{2} \times \frac{1}{2}$$

$$= \frac{1}{64}$$

Example 3

In a certain community $\frac{2}{3}$ of the residents are regular T.V. viewers. Fifty investigators are sent out to interview 5 people each.

a) Calculate the probability that 4 or more people out of the 5, are regular viewers.

b) How many investigators would you · expect to report that 4 or more people only viewed T.V. occasionally?

Solution

$$p = \frac{2}{3} = \text{regular T.V. viewers}$$

$$q = \frac{1}{3} = \text{non viewers or occasional viewers}$$

With 5 people, the expansion is $(p+q)^5 = 1$
From Pascal's Triangle, the coefficients are 1, 5, 10, 10, 5, 1.

Thus the expansion is

$(p+q)^5 = p^5 + 5p^4q + 10p^3q^2 + 10p^2q^3$
$\qquad\qquad\qquad\qquad\qquad + 5pq^4 + q^5$

a) Probability of 4 viewers $= 5p^4q = \frac{5}{1} \times \frac{2}{3} \times \frac{2}{3} \times \frac{2}{3} \times \frac{2}{3} \times \frac{1}{3} = \frac{80}{243}$

Probability of 5 viewers $= p^5 = \frac{2}{3} \times \frac{2}{3} \times \frac{2}{3} \times \frac{2}{3} \times \frac{2}{3} = \frac{32}{243}$

Probability of 4 or more viewers $= \frac{80}{243} + \frac{32}{243} = \frac{112}{243}$

b) Probability of 4 non-viewers $= 5pq^4 = \frac{5}{1} \times \frac{2}{3} \times \frac{1}{3} \times \frac{1}{3} \times \frac{1}{3} \times \frac{1}{3} = \frac{10}{243}$

Probability of 5 non-viewers $= q^5 = \frac{1}{3} \times \frac{1}{3} \times \frac{1}{3} \times \frac{1}{3} \times \frac{1}{3} = \frac{1}{243}$

Probability of 4 or more non-viewers $= \frac{10}{243} + \frac{1}{243} = \frac{11}{243}$

Expected no. of investigators $= \frac{11}{243} \times \frac{50}{1} = \frac{550}{243} = 2.3 = $ approx. 2.

Practical work

EXPERIMENT 1 *The frequency of heads when six coins are tossed together.*
For the experiment divide yourselves into groups of 6 pupils, each pupil having a coin to toss.

Copy this table into your jotter.

No. of trials = 64

No. of heads	Tally	Total
0		
1		
2		
3		
4		
5		
6		

All 6 members of the group should toss their coins at the same time and count up the number of heads showing, recording this in the table. Repeat this process until 64 trials in all have been completed.

Let us compare these experimental frequencies with the expected frequencies. Calculate the expected frequencies of 0, 1, 2, 3, 4, 5 and 6 heads, when 6 coins are tossed simultaneously 64 times, by using the binomial expansion of $(p+q)^6$.

Complete this table.

No. of heads	0	1	2	3	4	5	6
Experimental Frequency							
Theoretical Frequency							

Using this frequency table draw frequency polygons on the same graph for the experimental and theoretical frequencies of the number of heads showing when 6 coins are tossed together.

Compare your graph with the graphs of the other groups.

Make up another frequency table for experimental and theoretical results by totalling up the results of all the groups, and draw a graph as before. Compare this graph with your first one.

What conclusions do you reach about the distributions and in particular about the Symmetry of them?

EXPERIMENT 2 *The frequency of sixes when four dice are rolled together.*

This experiment may be done either by individuals or by groups of two or three pupils.

Copy this table into your book.

No. of trials = 72

No. of Sixes	Tally	Total
0		
1		
2		
3		
4		

Roll 4 dice together, 72 times in all, tallying the number of Sixes obtained each time and record the results in the table.

Now, calculate the expected frequencies of 0, 1, 2, 3, and 4 sixes when 4 dice are rolled together, 72 times, using the binomial expansion of $(p+q)^4$.

Fill in the results in the following table.

No. of Sixes	0	1	2	3	4
Experimental frequency					
Theoretical frequency					

Draw, on the same graph, frequency polygons for these experimental and theoretical frequencies of 0, 1, 2, 3, and 4 Sixes showing when 4 dice are rolled together. (In this graph consider your vertical scale carefully.)

Compare your graph with the graphs of the other members of the class.

As in Experiment 1, form another frequency table by adding up all the results for the class. Draw a graph for these frequencies. Compare this graph with your previous one.

What do you notice about the symmetry of this distribution? Consider the second graph of Experiment 1. What reasons can you give for points of similarity or difference?

Further experiments

Various other experiments may be devised and carried out along the same lines as the above experiments.

Several pairs of dice may be rolled at the same time.

Several sets of dice may be rolled and a coin may be tossed simultaneously.

A tetrahedron with its faces numbered may be rolled in the same way as a die.

Several pairs of tetrahedrons may be thrown at the same time.
etc.

EXERCISE A

1. On the average, a marksman firing at a target hits the bull's eye once in three shots. If he fires 6 times, what are the chances that he will hit the bull's eye
   a) twice *b*) 4 times *c*) not at all?

2. Another marksman, on average, hits the target 2 out of 3 times. In 4 shots what are his chances of hitting it 0, 1, 2, 3, or 4 times?

3. What is the probability of getting 4 heads in 8 tosses of a coin?

4. A shipment of oranges contains 10% bad ones. A random sample of 4 oranges is drawn from the shipment. Calculate the probability that the sample contains 0, 1, 2, 3 and 4 bad oranges.

5. In 10 throws of a coin what is the probability of obtaining 8 or more heads or tails?

6. In a certain large collection of plants, $\frac{2}{3}$ are single flowered varieties and the rest are double flowered. Calculate the probability of obtaining 0, 1, 2, 3, and 4 double flowered plants in a row of 4 plants.

7. Robert's chance of winning a set at tennis against David is $\frac{3}{4}$. Find his chance of winning at least 3 sets in a 5 set match assuming that all 5 sets are to be played.

8. An opinion poll finds that 3 out of 5 people are in favour of a certain proposal. What is the probability, that, if 3 people are taken at random, there will be a majority against the proposal?

9. In a certain town, the proportion of rainy to fine days in the month of June is 1 to 3. Assuming that each day is independant of the others, what is the chance that a week in June, in that town, will have
a) no wet days   *b*) 3 wet days?

10. From a box containing 6 white balls and 4 black balls, 3 balls are drawn at random. Find the probability that 2 are white and 1 is black.

11. In a certain industry, the workmen have a 20% chance of contracting an occupational disease. What is the probability that out of 6 workmen, 4 or more will contract the disease?

12. What is the probability of throwing at least 3 sevens in 5 throws of a pair of dice?

13. A coin is tossed 6 times. What is the probability of getting
   a) exactly 3 heads   *b*) at least 3 heads?

14. Seven dice are rolled. Calling a 5 or 6 a success, find the probability of having
   a) exactly 4 successes
   b) at most 4 successes

15. Assuming that the chance of a child being a boy or a girl is $\frac{1}{2}$, what is the probability that in a family of 6 there will be no fewer than 2 or more than 5 boys? What is the chance there will be 3 boys?

16. In a very large batch of screws, 5% are defective. What is the probability of finding at least 1 defective in samples of 5, 10, 15 and 20 respectively?

17. A manufacturer knows that an average 1 out of 10 of his products is defective. What is the probability that a random sample of 4 articles will contain
   a) no defectives
   b) exactly 1 defective
   c) at least 2 defectives
   d) no more than 3 defectives?

18. In a packet of flower seeds 40% are known to be pink flowering and the remainder white flowering. Calculate the probabilities of 0, 1, 2, 3, 4 or 5 pink flowers in a row of 5 plants. If 500 rows each of 5 plants are planted, approximately how many rows will contain
   a) all pink flowers
   b) all white flowers?

19. In a certain community, 2 out of every 3 houses has a telephone. If 50 investigators, each question 4 households, selected at random, how many may be expected to report that
   a) only 1 household out of the four does not have a telephone
   b) at least 2 out of the four households has a telephone?

20. One hundred interviewers are told to select 5 people at random and to question them concerning their drinking habits. Assuming that 3 out of 4 people drink alcohol at least occasionally, how many interviewers may be expected to find that
   a) at least 2 people are total abstainers.
   b) at least 3 people drink alcohol sometimes?

11 THE NORMAL DISTRIBUTION

Frequency distributions

Let us consider the following frequency distributions where the total frequency is fairly high.

Example 1
This distribution shows the number of children per family in the families of all first year pupils in a Secondary School in a country area. (This cannot be said to be quite typical of the families in the area as a whole since families with no children are not considered.)

Figure 11.1 shows the histogram for this distribution.

Number in family	1	2	3	4	5	6	7	8	9	10	11	12	13
Frequency	10	27	37	36	22	18	14	9	4	3	0	2	1

Fig. 11.1 Number of Children per Family

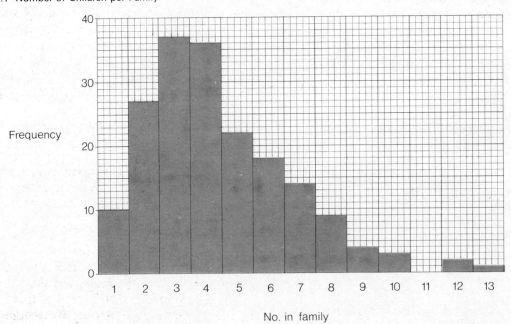

Frequency

No. in family

As you can see from the histogram, this is a rather 'lopsided' distribution. This kind of distribution is said to be **skewed.**

Example 2
The results recorded in a cycling proficiency test given by the police to school children in Wigtownshire are shown below. This was not a compulsory test, pupils could enter or not as they wished. Most of the children were aged about 10–12 years.

Marks	Frequency
35–39	1
40–44	0
45–49	0
50–54	1
55–59	4
60–64	3
65–69	6
70–74	16
75–79	22
80–84	75
85–89	113
90–94	68
95–99	5

Draw a histogram to illustrate this. Can you suggest any reasons for the shape of the distribution? Is this distribution skewed? What shape do you think it might have had if every child in the area aged 10–12 years had entered the test?

Example 3
These deaths (from all causes) were recorded one year in Scotland.

Age (years)	Frequency
0– 4	1619
5– 9	117
10–14	91
15–19	177
20–24	177
25–29	175
30–34	208

Age (years)	Frequency
35–39	392
40–44	682
45–49	960
50–54	1846
55–59	2951
60–64	4165
65–69	4181
70–74	4343
75–79	4082
80–84	3182
85–89	1814
90–94	523
95–99	73
100 +	5

Without any further grouping, draw a histogram for this distribution and discuss the shape of it.

This shape of distribution is the usual one for deaths and it is an example of what is called a U type distribution.

Example 4

A ten sum test for speed and accuracy was given to all first year pupils in a country area and the results were as follows.

Number of sums correct	0	1	2	3	4	5	6	7	8	9	10
Frequency		5	11	21	35	49	52	51	28	20	13	3

Figure 11.2 (a) (on the next page) shows the histogram for this

This histogram is fairly symmetrical, as you might expect since it represents the complete range of abilities in this area.

The mean, median and mode are all 5.

In this chapter we are principally interested in distributions of this type, i.e. symmetrical about the mode and shaped like a bell. The standard distribution of this type is called a **Gaussian** or **normal distribution.**

101

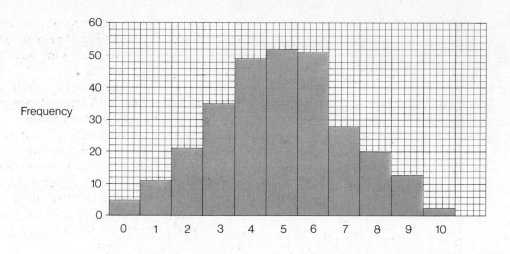

No. of sums correct

Fig. 11.2 Sum Test

Example 5

Some of the first year pupils who sat the above test, had already done this same test several months before when they were at the end of their primary education.

Number of sums correct	0	1	2	3	4	5	6	7	8	9	10
Frequency (Primary)	0	1	6	7	15	23	20	19	22	8	8
Frequency (Secondary)	3	2	9	16	20	23	18	18	11	7	2

On the same graph, draw frequency polygons for the two distributions, primary and secondary. Comment on the shape of the distributions and give reasons for the difference in the results from the one test to the next.

Note: The sums given in the test are shown below. The pupils sitting the test were allowed to write their names on the testpaper and then given *exactly* $3\frac{1}{2}$ minutes to answer it.

TEST PAPER

1. Add

```
  55
 346
 107
 432
```

2. Add

```
5038
 967
2864
  89
4692
```

3. Add

```
6875
 428
5069
3786
  53
8647
```

4. Subtract

```
7404
3856
```

5. Subtract

```
25983
 9087
```

6. Subtract

```
70136
49768
```

7. Multiply

```
 276
× 48
```

8. Multiply

```
5397
× 76
```

9. Divide

```
9)8658
```

10. Divide

```
26)13962
```

102

Practical work

Collect some data of your own by undertaking some surveys. There are two principal types of surveys that may be conducted—those dealing with data about objects and those dealing with information about people.

Here are a few suggestions that you may care to follow up. A little thought and some help with apparatus from perhaps the science department, should enable them to be carried out fairly easily.

In each case a reasonably large number of observations should be taken and a frequency table constructed. Then the mean and standard deviation could be calculated and the results shown graphically as a histogram or frequency polygon.

Suggestions

1. An investigation could be conducted into the length of life of torch batteries. Here you would need to borrow some electrical equipment from the science department for the setting up of electrical circuits.

2. The height of bounce of a large number of tennis (or golf) balls could be measured. In this experiment, the balls must be dropped from a fixed height and the heights of the resulting bounce judged against a background board marked off in suitable measuring units. Obviously this work requires a team of pupils working together.

3. An investigation into the weight range of 'standard' eggs could be very interesting. You should be able to 'borrow' a large number of 'standard' eggs from the school kitchen or the cookery teacher and weigh them accurately with a science balance. To obtain enough weighings, the experiment could be spread over several weeks. Once you have formed a frequency distribution you will know the range of weights which covers the category of 'standard' eggs. Then you can contact an egg packing station and find out what the official range of weights is, and compare this with your own results.

4. A similar type of experiment may be done by very accurate weighing of packets of salt, custard powder, sugar etc. which have a stated weight on the packet.

5. A survey of the heights of a certain year group or age group in the school may be conducted quite readily (with some cooperation from the staff). All that is required is some kind of measuring instrument—at its simplest a metre stick tacked onto a wall. Here of course, boys and girls should be considered separately.

6. The same type of survey may be done by taking some other measurement of a group in the school, e.g. weights.

These last two surveys can be broadened in scope by contacting another school or several schools in various parts of the country and 'swopping' information with them. This would give several distributions for purposes of comparison. It would also be interesting to keep a record of things like heights and weights in a school over a number of years.

7. Many other surveys may be conducted in a school, e.g. sizes of shoes, the number of children per family, speed and accuracy sum tests, etc.

The normal distribution

In Example 4, at the beginning of the chapter, we saw a distribution symmetrical about the mode (and the mean and median) which approximated to a **normal distribution.** Figure 11.3 shows again the histogram of this distribution. This time a curve has been drawn through the tops of the columns. This is roughly the typical bell shaped curve of a normal distribution and is called the **normal curve.**

The normal curve

The algebraic study of this curve took place in the middle of the eighteenth century and is mainly attributed to De Moivre and Gauss. Their aim was to obtain the equation of a curve which corresponded to the

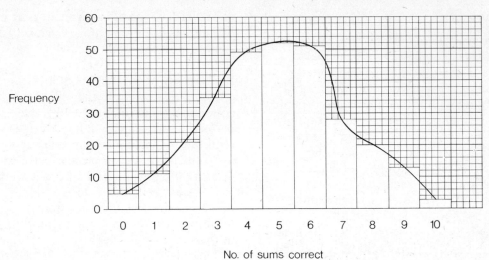

Fig. 11.3 Sum Test

symmetrical bell shape. In the distribution the mean, median and mode coincide and the relative width of the 'bell' will depend on the spread of the distribution.

The details of this study are beyond the scope of this book but it is worth noting that the equation that is used in relation to the normal distribution curve is

$$y = \frac{N}{\sigma\sqrt{2\pi}}e^{\frac{-(x-\mu)^2}{2\sigma^2}}$$

where x is the continuous variable,
     N is the total number of observations
     μ is the mean of the distribution,
and σ is the standard deviation of the distribution.

The size and shape of the "bell" depends on the scales used.
In b) the horizontal scale has been shortened and in c) the vertical scale has been lengthened.

Fig. 11.5 The Size and Shape of the 'Bell' depend on the Scales Used

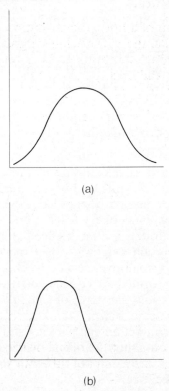

(a)

(b)

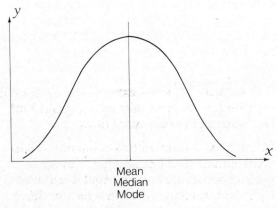

Mean
Median
Mode

Fig. 11.4 The Normal Curve

104

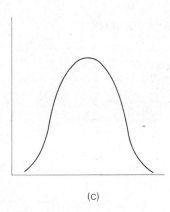

(c)

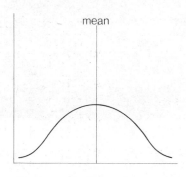

A distribution which is shaped like the normal curve is called a **normal distribution.** A great many distributions which occur naturally are normal distributions e.g. the weights of a certain age group of men, the number of tomatoes per plant, the number of peas in a pod. Your practical work should have produced normal distributions also.

The mean and standard deviation of the normal curve

The mean of a normal distribution lies in the middle, the curve being symmetrical. The standard deviation gives us the measure of the spread or dispersion of the observations.

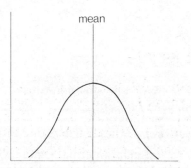

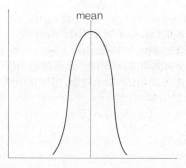

Fig. 11.7

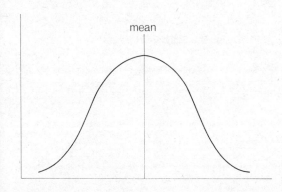

Fig. 11.6

Figure 11.7 shows 3 normal distributions with the same mean but differing standard deviations.

It is clear that when examining any normal distribution we must take into account both the mean and the standard deviation.

In this way, we are able to compare normal distributions with one another, though in practice we compare all normal distributions with the special distribution which we have already called the Normal Curve.

105

Standard normal curve

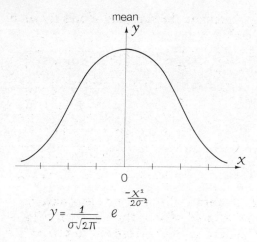

$$y = \frac{1}{\sigma\sqrt{2\pi}} \, e^{\frac{-X^2}{2\sigma^2}}$$

Fig. 11.8 Standard Normal Curve

Since the shape and position of each normal curve depends upon the mean and standard deviation of the distribution it will simplify our work if we standardise our data. Thus we obtain the **standard normal curve** which is a special version of the Normal Curve with the mean $(\mu) = 0$, $N = 1$, and the standard deviation $(\sigma) = 1$. Further, the area between the curve and the x-axis is 1 unit and thus the area between the curve, the x-axis the ordinate $x = a$ and the ordinate $x = b$ is a measure of the probability that the variable x lies between $x = a$ and $x = b$. These probabilities have all been accurately calculated and are usually tabulated in standard sets of tables. Some examples of these probabilities are illustrated below.

Fig. 11.9

Standardising scores

A point x on a normal curve with mean $= \mu$ and standard deviation $= \sigma$ corresponds to the point Z on the standard normal curve when $Z = \dfrac{x - \mu}{\sigma}$

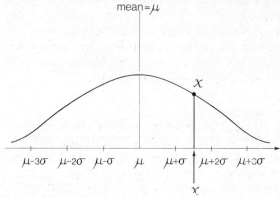

Fig. 11.10 (a) A Normal Curve

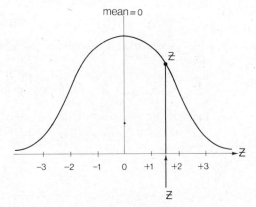

Fig. 11.10 (b) The Standard Normal Curve

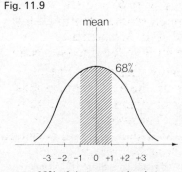

68% of the area under the curve lies with 1 Standard Deviation of the Mean.

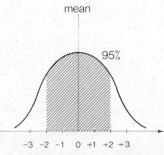

95% of the Area under the curve lies within 2 Standard Deviations of the Mean

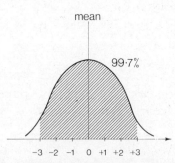

99.7% of the area under the curve lies within 3 Standard Deviations of the Mean

106

Thus we can calculate the corresponding standard score (Z) for any score (x) in a normal distribution by using the formula

$$\text{Z (Standard Score)} = \frac{x - \mu}{\sigma}$$

(This process is called standardising the score.)
In this way, we can compare scores in different normal distributions by finding the standard score for each and seeing how they stand in relation to one another.

Example
In a statistics examination the mean mark was 78.0 and the standard deviation 8.0. In algebra, the mean was 73.0 and the standard deviation 7.6. A pupil scored 75 in statistics and 71 in algebra. In which examination was his relative standing higher?

Solution

Statistics: $\mu = 78.0$ Algebra: $\mu = 73.0$
$\quad\quad\quad\;\; \sigma = 8.0$ $\quad\quad\quad\;\; \sigma = 7.6$
$\quad\quad\quad\;\; x = 75$ $\quad\quad\quad\;\; x = 71$

$$Z = \frac{x - \mu}{\sigma} \quad\quad\quad Z = \frac{x - \mu}{\sigma}$$

$$= \frac{75 - 78}{8} \quad\quad\quad = \frac{71 - 73}{7.6}$$

$$= \frac{-3}{8} \quad\quad\quad\quad = \frac{-2}{7.6}$$

$$= -0.375 \quad\quad\quad = -0.263$$

i.e. the standard score   i.e. the standard score
for statistics $= -0.375$ for algebra $= -0.263$
(0.375   (0.263
below   below
the   the
mean)   mean)

Relatively speaking, the pupil did better in algebra than in statistics.

Note: It is of benefit in working this type of example to summarise the information given about each subject before going on to calculate the standard score.

EXERCISE A

1. Calculate the standard scores for a), b), c), d).

	a)	b)	c)	d)
x	72	45	63	71
μ	60	50	60	79
σ	9	5	11	5

2. A normal distribution has a mean $\mu = 120$ and a standard deviation $\sigma = 11$. What standard scores correspond to raw scores of a) 115, b) 134, c) 93?

3. Mary had a Geography mark of 63 in one examination where the average mark was 70 and the standard deviation 7. In the next examination, where the average was 71 she scored 65 (standard deviation being 5). Compare her standing in the two examinations.

4. John scored 75 in an arithmetic test where the mean mark was 70 and the standard deviation 6. His sister Jane scored 70 in her arithmetic test, the mean and standard deviation being 60 and 8 respectively. Which of two showed more ability in Arithmetic?

5. In an English examination the mean mark was 56 and the standard deviation was 8; in French the mean was 60 and standard deviation 14; in Mathematics, the mean and standard deviation were 65 and 10 respectively. A pupil scored 60 in English, 67 in French and 70 in Mathematics. Compare his standing in the three subjects.

6. Tom and Bob are friends who attend different schools. In the term examinations Tom had a mark of 70 in English and Bob had a mark of 80. The average mark in Tom's class was 65 with a standard deviation of 11. In Bob's class the average was 70 and the standard deviation was 10. Which of the boys had actually achieved the better result?

7. A pupil scored 50 in English and 62 in Mathematics in the November examination, the average marks being 55 and 60 respectively. In May, he scored 55 in English and 60 in Mathematics, the average mark being 58 for both subjects. The standard deviation for English was 7 in November and 5 in May; and for Mathematics was 5 in November and 12 in May.
   a) In which examination did he do better in English?
   b) In which examination did he do better in Mathematics?
   c) In which of all 4 examinations had he the best results?
   d) In which of all 4 examinations had he the poorest results?

8. In a normal distribution with mean 30 and standard deviation 3, what raw scores correspond to the following standard scores?
   a) 1.4   b) 2.5   c) −1.5   d) −0.68

9. In a normal distribution with mean 55 the raw score 60 corresponds to a standard score of 1.3. What is the standard deviation of the distribution (correct to the second decimal place)?

10. A normal distribution has a mean of 72, and a raw score of 60 corresponds to a standard score of −1.2. What is the standard deviation?

11. What is the mean of a normal distribution which has a standard deviation of 5 and in which a raw score of 30 corresponds to a standard score of 1.7?

12. In a normal distribution, the standard deviation is 2.3 and a score of 62 corresponds to a standard score of −0.5. What is the mean of the distribution?

13. What are the mean and the standard deviation of a normal distribution when raw scores of 45 and 60 correspond to standard scores of −1 and 2 respectively?

14. In a normal distribution scores of 70 and 50 correspond to standard scores of 0.5 and −1.5 respectively. Calculate the mean and standard deviation of the distribution.

108

Area under the standard normal curve

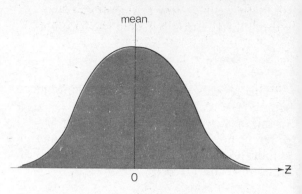

Fig. 11.11 (a) The total area under the Standard Normal Curve equals 1 (or 100%)

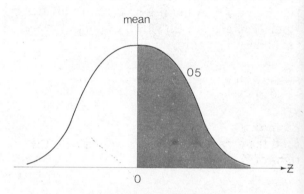

Fig. 11.11 (b) The area to the right of the mean (above the mean) equals 0.5 (or 50%). Similarly the area to the left of the mean (below the mean) equals 0.5 (or 50%)

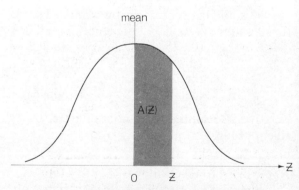

Fig. 11.11 (c) The shaded area under the curve A(Z) is the proportion of area between Z = 0 and any value Z

Tables have been compiled whereby we can read off the size of part of the area under the Standard Normal Curve provided we know the standard score.

Example 1
Find the area under the normal curve between $Z = 0$ and $Z = 1.2$.

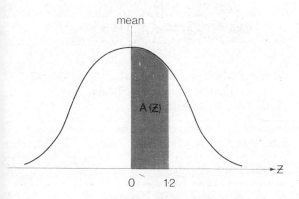

Fig. 11.12 (a) $Z = 1.2$ therefore $A(Z) = 0.385$

Example 2
Find the area under the normal curve to the right of $Z = 1.2$.

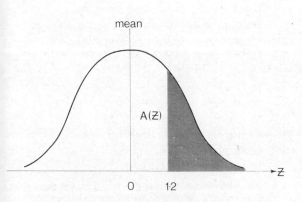

Fig. 11.12 (b) $Z = 1.2$ therefore $A(Z) = 0.385$
Required area $= 0.5 - 0.385 = 0.115$

Note: A (Z) is the area from the mean to Z. Since we require the area to the right of $Z = 1.2$, we must subtract $A(Z)$ from the total area to the right of the mean which is 0.5.

Example 3
Find the area between $Z = 0$ and $Z = -1.2$

$$Z = -1.2$$
$$\therefore A(Z) = 0.385$$

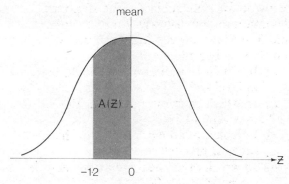

Fig. 11.13

(In the tables, only positive values of Z are given. Since the curve is symmetrical, the area for $Z = -1.2$ is the same as the area for $Z = +1.2$.)

Example 4
What is the area between $Z = 0.81$ and $Z = 1.94$?

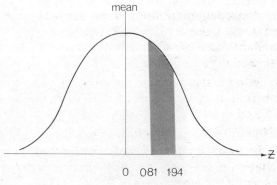

Fig. 11.14

$$Z_1 = 1.94$$
$$\therefore A(Z_1) = 0.474$$
$$Z_2 = 0.81$$
$$\therefore A(Z_2) = 0.291$$
$$\text{Required area} = 0.474 - 0.291$$
$$= 0.183$$

109

Example 5
Calculate the area from $Z = -0.46$ to $Z = 1.21$

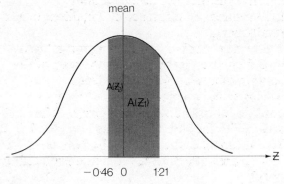

Fig. 11.15

$$Z_1 = 1.21$$
$$\therefore A(Z_1) = 0.387$$
$$Z_2 = -0.46$$
$$\therefore A(Z_2) = 0.177$$
$$\text{Required area} = 0.387 + 0.177$$
$$= 0.564$$

Example 6
A normal distribution has mean $\mu = 10$ and standard deviation $\sigma = 2$. Find the area under the curve from $x = 8$ to $x = 13$.

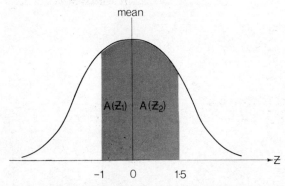

Fig. 11.16

First of all, we must standardise the scores.

$$x_1 = 8$$
$$Z_1 = \frac{x_1 - \mu}{\sigma} = \frac{8 - 10}{2} = \frac{-2}{2} = -1$$

$$x_2 = 13$$
$$Z_2 = \frac{x_2 - \mu}{\sigma} = \frac{13 - 10}{2} = \frac{3}{2} = 1.5$$
$$Z_1 = -1$$
$$\therefore A(Z_1) = 0.341$$
$$Z_2 = 1.5$$
$$\therefore A(Z_2) = 0.433$$
$$\therefore \text{Required area} = 0.341 + 0.433$$
$$= 0.774$$

Note: A diagram, as shown in the opposite column, for each example is of great benefit.

EXERCISE B

Find the area under the normal curve in each of the following cases.
1. from $Z = 0$ to $Z = 1.3$
2. from $Z = 0$ to $Z = 2.43$
3. from $Z = 0$ to $Z = -0.6$
4. from $Z = 0$ to $Z = -1.05$
5. from $Z = -0.6$ to $Z = 1.23$
6. from $Z = -1.78$ to $Z = 2.36$
7. from $Z = 0.4$ to $Z = 1.7$
8. from $Z = -0.62$ to $Z = -2.33$
9. from $Z = 0.1$ to $Z = 0.33$
10. from $Z = -1.6$ to $Z = -0.01$
11. to the right of $Z = 1.2$
12. to the right of $Z = 3.24$
13. to the left of $Z = -0.62$
14. to the left of $Z = -1.4$
15. to the right of $Z = -2.11$
16. to the right of $Z = -0.5$
17. to the left of $Z = 1.32$
18. to the left of $Z = 0.34$
19. for values of Z greater than 0.78
20. for values of Z greater than -0.78
21. for values of Z less than 1.30
22. for values of Z less than -0.88
23. A normal distribution has mean $\mu = 12$ and standard deviation $\sigma = 2$. Find the following areas under the normal curve.
   a) from $x = 10$ to $x = 13.5$
   b) from $x = 12.5$ to $x = 14$
   c) from $x = 8$ to $x = 9.6$

d) for values of x exceeding 14

e) for values of x less than 11

24. The following areas refer to areas under the standard normal curve. From the information given find the values of Z (use the Tables of Area).

   a) The area between 0 and Z is 0.379

   b) The area between 0 and Z is 0.486

   c) The area to the left of Z is 0.323

   d) The area to the right of Z is 0.192

   e) The area to the right of Z is 0.595

   f) The area to the left of Z is 0.9

   g) The area between 0 and Z is 0.493

   h) The area between 0 and Z is 0.497

   i) The area between 0 and Z is 0.490

Probability and area

Probability and the areas of histograms or distribution graphs are very closely connected. Let us consider this frequency distribution of the number of children per family in 50 families

No. of children	0	1	2	3	4	5	6	7
Frequency	3	6	12	15	8	3	2	1

The number of families out of the 50 with 3 children was 15. Thus the probability of choosing a family, at random, with 3 children in it is $\frac{15}{50}$.

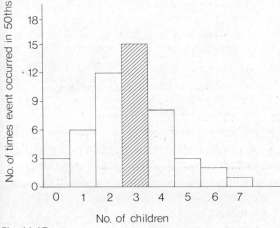

Fig. 11.17

The scale of the histogram may be chosen so that the total area enclosed by the columns is 1. To do so in this case, the vertical scale is in 50ths. Thus the area of the column representing the incidence of 3 children in a family is $\frac{15}{50}$.

So, the probability of picking out a family with 3 children is the same as the area of the column representing that incidence.

This same principle applies when we consider normal distribution curves. The area under the standard normal curve is 1. This means that the total probability is 1; and an area of 0.3 is the same as a probability of 0.3.

Example

In a normal distribution with mean 10 and standard deviation 2, what is the probability that x is greater than 12?

Solution

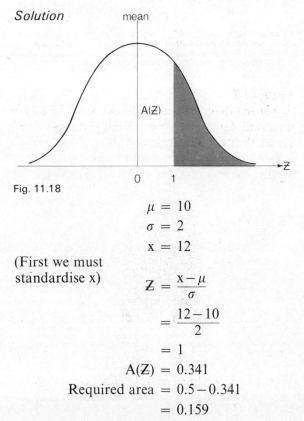

Fig. 11.18

$$\mu = 10$$
$$\sigma = 2$$
$$x = 12$$

(First we must standardise x)

$$Z = \frac{x - \mu}{\sigma}$$
$$= \frac{12 - 10}{2}$$
$$= 1$$
$$A(Z) = 0.341$$
$$\text{Required area} = 0.5 - 0.341$$
$$= 0.159$$

i.e. the probability of x being greater than 12 is 0.159.

EXERCISE C

1. Given that x is normally distributed with mean 50 and standard deviation 5, use the tables to calculate the probability that

a) x>60   b) x>55   c) x<43

d) x<40   e) 40<x<60

2. Assuming that the stature (x) of female university students is normally distributed with mean 160 cm and standard deviation 5 cm, calculate the probability that

a) x>152.5 cm

b) 150<x<165

The normal curve as a limit of a frequency distribution of a continuous variable

The normal curve serves as a good approximation for the histograms obtained from many distributions of continuous variables.

Example 1
A brand of electric light bulbs has an average life of 1 year and a standard deviation of 3 months. What is the probability that a bulb chosen at random will have a life of at least 18 months?

Solution

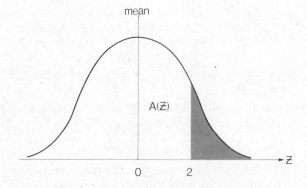

Fig. 11.19

112

$\mu = 12$ months

$\sigma = 3$ months

$x = 18$ months

$$Z = \frac{x - \mu}{\sigma}$$

$$= \frac{18 - 12}{3}$$

$$= \frac{6}{3}$$

$$= 2$$

$$\therefore A(Z) = 0.477$$

$$\text{Required area} = 0.5 - 0.477$$

$$= 0.023$$

i.e. the probability of the bulb having a life of at least 18 months (18 months or more) is 0.023.

Example 2
Sacks of grain packed by an automatic machine loader have an average weight of 114 kg. It is found that 10% of the bags are over 116 kg. Find the standard deviation.

Solution

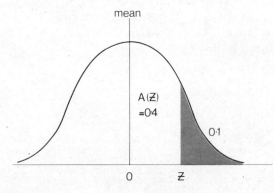

Fig. 11.20

In this type of example we must start with the area under the curve and work back to Z.

We know 10% of the bags are over 116 kg. This means the area to the right of Z is 10% or 0.1.

$$A(Z) = 0.5 - 0.1$$
$$= 0.4$$
$$\therefore Z = 1.28$$

Now we know, $\mu = 114\,kg.$

$$\sigma = ?$$
$$x = 116\,kg.$$
$$Z = 1.28$$
$$Z = \frac{x - \mu}{\sigma}$$
$$\frac{1.28}{1} = \frac{116 - 114}{\sigma}$$
$$= \frac{2}{\sigma}$$
$$1.28\sigma = 2$$
$$\sigma = \frac{2}{1.28}$$
$$= 1.56 \text{ (correct to 2nd decimal place)}$$

i.e. the standard deviation is 1.56

Example 3
Given a normal distribution of a continuous variable (x) with 2000 variates, the mean and standard deviation being 20 and 5 respectively, find the number of variates between x = 12 and x = 22.

Solution

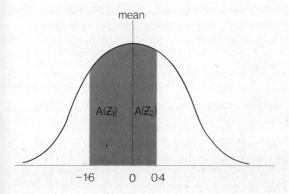

Fig. 11.21

$$N = 2000$$
$$\mu = 20$$
$$\sigma = 5$$
$$x_1 = 12$$
$$x_2 = 22$$
$$Z_1 = \frac{x_1 - \mu}{\sigma}$$
$$= \frac{12 - 20}{5}$$
$$= -1.6$$
$$Z_2 = \frac{x_2 - \mu}{\sigma}$$
$$= \frac{22 - 20}{5}$$
$$= 0.4$$
$$\therefore A(Z_1) = 0.445$$
$$\text{and } A(Z_2) = 0.155$$
$$\therefore \text{Required area} = 0.445 + 0.155$$
$$= 0.600$$

i.e. the probability of x lying between 12 and 22 is 0.6
∴ The expected number of variates between 12 and 22 = 2000×0.6

$$= 1200$$

EXERCISE D

1. A hundred standard squash balls are tested by dropping from a height of 250 cm and measuring the height of bounce. A ball is 'fast' if it rises above 80 cm. The average height of bounce was 75 cm and the standard deviation was 1.9 cm. What is the chance of getting a 'fast' standard ball?
2. Sacks of grain packed by an automatic loader have an average weight of 61 kg. It is found that 10% of the bags are below 59 kg. Find the standard deviation.
3. In the previous example, the machine is adjusted and the average weight per bag is now 60 kg. Assuming the standard deviation

is unaltered calculate the probability that a bag is now below 59 kg.

4. Suppose your score in an examination in standard units is 1.2 and the scores are assumed to be normally distributed, what percentage of the students would be expected to score higher than you?

5. Two brands of torch batteries have the same average life of 55 hours but different standard deviations of $1\frac{1}{4}$ hours and $1\frac{3}{4}$ hours. In each case, what is the chance that a battery will burn at least 58 hours?

6. A physical education teacher gives grades relative to all his classes for athletic events. If experience has shown that the average height is 178 cm and the standard deviation is 10 cm in the high jump, and the teacher gives 20% A's, how high would a pupil need to jump to expect an A?

7. If a normal distribution of a continuous variable has mean 21.2 and standard deviation 3.10, find the probability that a variate selected at random will be larger than 30 or less than 15.

8. The weight of grapefruit from a large shipment averages 420 g with a standard deviation of 42 g. If these weights are normally distributed, what percentage of all these grapefruit would be expected to weigh between 420 and 476 g?

9. If the average life of a certain make of storage battery is 30 months with a standard deviation of 5 months, what percentage of these batteries can be expected to last 24 to 36 months?

10. Records indicate that the average life of T.V. tubes is 3 years 6 months and the standard deviation is 18 months. Tubes lasting less than a year are replaced free. For every 200 sets (one tube to a set) sold, how many tubes can be expected to have to be replaced free?

11. The average yearly rainfall in a certain town is 75 cm with a standard deviation of 25 cm. Assuming this yearly rainfall follows a normal distribution, in how many years out of a period of 100 years could the residents of the town expect less than 37.5 cm of rain?

12. If the heights of 1000 male college students closely follow a normal distribution with a mean of 172.5 cm and a standard deviation of 6.25 cm.
   a) how many of these students would you expect to be at least 180 cm tall?
   b) what range of heights would include the middle 50% of the men in this group?

13. In grading a certain type of plum, 20% are called small, 55% medium, 15% large and 10% very large. If the weights follow a normal distribution, and the average weight is 138 g with a standard deviation of 33.5 g, what are the lower and upper bounds for the weight of medium plums?

14. In a normal distribution with a standard deviation of 2, the probability that a variate selected at random exceeds 26 is 0.05. Find the mean of the distribution and the variate above which lie 80% of all variates of the distribution.

15. In an examination, the average mark was 60 and the standard deviation 10. The teacher gave 30 students with marks between 51 and 69 a grade of C. If the marks are assumed to follow a normal distribution, how many students sat the examination?

16. A normal distribution has a mean of 30 and a standard deviation of 7. If it is known that 346 variates exceed 32.5, what is the total number of variates in the distribution?

17. A teacher who gives 10% A's, 20% B's, 40% C's, 20% D's and 10% E's, sets an examination in which the average mark is 54. If the borderline between the C's and the

D's is 44, what is the standard deviation in that examination?

18. In a normal distribution with mean 100 and standard deviation 44, there are 150 variates greater than 180. How many variates should we expect between 120 and 180?

19. In a population of adult men 20% are 175 cm or over in height and 10% are less than 155 cm. Find the mean and standard deviation of the population.

20. In an examination 10% of the class receive a grade of A, 20% B, 55% C and 15% D. The C grade ranges from 55 to 70. Assuming a normal distribution, what are the mean and standard deviation of the marks?

The binomial distribution as an approximation to the normal distribution

If a frequency distribution is obtained by using the terms of the binomial expansion, it is called a binomial distribution, e.g. the expected frequencies of the occurrence of heads in tossing N coins follow a binomial distribution.

The histograms representing the expected frequencies of heads in 64 tosses of 1, 2, 3, 4, 5 and 6 coins are shown below.

If the number of the sample, N, is large the calculation of frequencies and probabilities by means of the binomial theorem becomes tedious. Since many practical problems involve samples of large size, it is important to find a more rapid method of calculating probabilities. Such a method is furnished by the normal distribution which is the most important continuous probability distribution.

From the figures in Figure 11.22 it is apparent as N increases, the tops of the rectangles of the histograms approach a bellshaped curve. This limiting frequency curve obtained as N becomes larger and larger is called the normal frequency curve (or normal probability curve). The curve approaches the

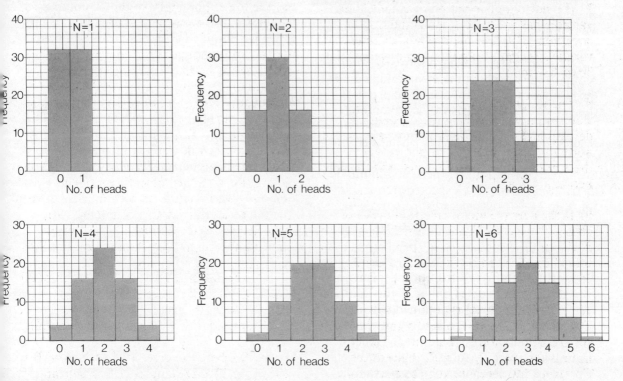

Fig. 11.22

115

horizontal axis but never touches it. The total area under the normal probability curve representing the probability that x will fall anywhere is equal to 1.

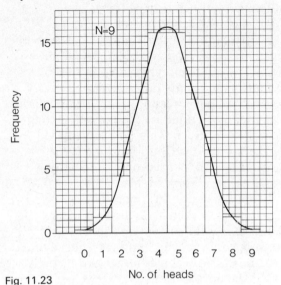

Fig. 11.23

Hence, by using the tables for finding the areas under the standard normal curve, we can find the probability of a certain event occurring.

To calculate the standard score (or Z score), we must know the mean and standard deviation of the distribution. The proof of the values of these measures is beyond our scope and here we shall only quote the results. If p is the probability of the success of an event in a *single trial* and q is the probability of its failure, then the binomial distribution giving the expected frequencies for 0, 1, 2 ------ N successes in N trials has

$$\text{mean } \mu = Np$$

and standard deviation $\sigma = \sqrt{Npq}$

Note : The binomial distribution deals with *discrete* variables only e.g. the number of times a head is obtained when tossing a coin, the number of times a 6 is obtained with a die, the number of defective T.V. tubes or light bulbs etc. This means that in working examples, if we wish to include a certain score we have to adjust the x value used.

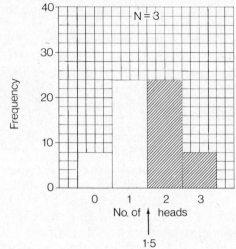

Fig. 11.24

Let us look at the third histogram shown in Figure 11.22 of the expected number of heads obtained in 64 tosses of 3 coins. If we wish to know how many times we obtained 2 heads or more, then to include 2 heads we must use the lower boundary of the '2' column which is 1.5. If we want to know how many times we obtained 1 or less heads we must use the upper boundary of the '1' col. which is 1.5.

Example 1
Using the normal curve approximation, find the probability that 3–6 heads are obtained in a toss of 9 coins (i.e. the probability that x lies between and *includes* 3 and 6).

Solution

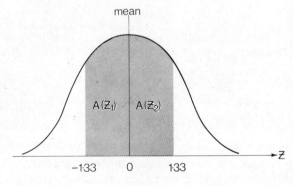

Fig. 11.25

$p = \frac{1}{2}$ = heads

$q = \frac{1}{2}$ = tails

$N = 9$

$\mu = Np$

$ = 9 \times \frac{1}{2}$

$ = 4.5$

$\sigma = \sqrt{Npq}$

$ = \sqrt{9 \times \frac{1}{2} \times \frac{1}{2}}$

$ = 1.5$

$x_1 = 2.5$ (to include 3 heads)

$x_2 = 6.5$ (to include 6 heads)

$Z_1 = \dfrac{x_1 - \mu}{\sigma}$

$ = \dfrac{2.5 - 4.5}{1.5}$

$ = \dfrac{-2}{1.5}$

$ = -1.33$

$A(Z_1) = 0.408$

$Z_2 = \dfrac{x_2 - \mu}{\sigma}$

$ = \dfrac{6.5 - 4.5}{1.5}$

$ = \dfrac{2}{1.5}$

$ = 1.33$

$A(Z_2) = 0.408$

Required area $= 0.408 + 0.408$

$ = 0.816$

i.e. the probability of obtaining 3—6 heads is 0.816.

Example 2
If 12 dice are thrown, what is the probability, using the normal curve approximation, that 6 or more dice will show a 5?

Solution

$p = \frac{1}{6}$ = a 5

$q = \frac{5}{6}$ = any other number.

$N = 12$

$\mu = Np$

$ = 12 \times \frac{1}{6}$

$ = 2$

$\sigma = \sqrt{Npq}$

$ = \sqrt{12 \times \frac{1}{6} \times \frac{5}{6}}$

$ = 1.29$

$x = 5.5$ (to include 6 fives)

$Z_1 = \dfrac{x - \mu}{\sigma}$

$ = \dfrac{5.5 - 2}{1.29}$

$ = \dfrac{3.5}{1.29}$

$ = 2.71$

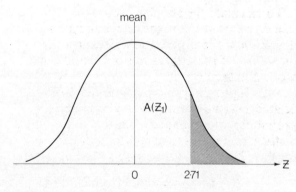

Fig. 11.26

$A(Z_1) = 0.497$

Required area $= 0.5 - 0.497$

$ = 0.003$

i.e. the probability of obtaining 6 or more fives is 0.003.

117

Example 3

A manufacturer producing tea cups finds that 10% of them are defective. What is the probability that in a sample of 50 cups, no more than 4 are defective?

Solution

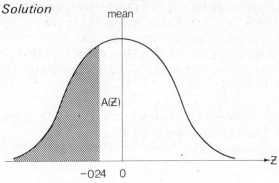

Fig. 11.27

$$p = \frac{1}{10} = \text{defective}$$

$$q = \frac{9}{10} = \text{not defective}$$

$$N = 50$$

$$\mu = Np$$

$$= 50 \times \frac{1}{10}$$

$$= 5$$

$$\sigma = \sqrt{Npq}$$

$$= \sqrt{50 \times \frac{1}{10} \times \frac{9}{10}}$$

$$= 2.12$$

$$x = 4.5 \text{ (to include 4)}$$

$$Z = \frac{x - \mu}{\sigma}$$

$$= \frac{4.5 - 5}{2.12}$$

$$= \frac{-0.5}{2.12}$$

$$= -0.24$$

$$A(Z) = 0.095$$

Required area $= 0.5 - 0.095$

$$= 0.405$$

i.e. the probability of obtaining no more than 4 (4 or less) defectives is 0.405.

EXERCISE E

1. A coin is tossed 100 times. Use the normal curve approximation to find the probability of obtaining
    a) 60 or more heads
    b) exactly 50 heads

2. Use the normal curve approximation to find the probability of obtaining exactly 16 sixes in 96 tosses of a die.

3. For a binomial frequency for which $p = \frac{1}{4}$ find the probability of obtaining 25 or more successes in 80 trials.

4. If 40% of students have defective eyesight, what is the probability that at least half of the members of a class of 50 students will have defective eyesight? Use the normal curve approximation.

5. If 10% of T.V. tubes burn out before their guarantee has expired,
    a) what is the probability that a shopkeeper who has sold 200 such tubes will have to replace at least 30 of them?
    b) what is the probability that he will replace at least 10 and not more than 30 tubes.
Use the normal curve approximation.

6. If 25% of the drivers in a certain town have at least one accident in a year's driving, what is the probability that 30% or more of 500 customers of an insurance company will have an accident during the next year? Use the normal curve approximation.

7. Assume that half of the people in a certain community are regular viewers of T.V. Of 100 investigators, each interviewing 10

118

individuals regarding their viewing habits, how many would you expect to report that 3 people or fewer were regular T.V. viewers? (Use normal distribution.)

8. A manufacturer of light bulbs finds that on average 5% are defective. What is the probability that out of 1000 such bulbs selected at random 30 or more are defective?

9. A pair of dice is rolled 25 times. What is the probability that a 7 will show 10 or more times?

10. According to a certain insurance company, the probability of a man aged 45 dying within a year is 0.05. If the company has 20 000 policies in force on men of this age estimate the probability that the company will have to pay out more than 900 death claims on this group of men within a year.

11. A manufacturer of radio valves knows that on average 2% of his products are defective. Using the normal curve approximation, what is the probability that a sample of 100 valves will contain exactly 5 defective valves?

12. How many times would 2 dice have to be rolled so that there is a 50% chance of getting at least 3 fives?

12 SAMPLING

Have you ever sampled your mother's baking? Presumably you did not eat all of a batch of cakes to see what the baking was like. You tasted one cake and assumed that they were all of the same quality. But what if the one you tasted was too near the heat in the oven and was burned? Was this typical of all the cakes or was it an exception? This is the kind of thing we must consider when we take a sample of something.

Sampling things is part and parcel of everyday life. A farmer will run his hand through a sack of grain and pick out a handful to examine. A grocer may give a woman a small piece of cheese to taste so that she can decide whether to buy some or not. Often manufacturers of breakfast cereals, soap powders, etc, will distribute small sample boxes of their products to housewives so that they can try them.

If you stop to think about it, you will find that almost all the facts we know have been discovered by sampling. We know what rivers are like because we have seen a few, not all of them. We know what poetry is like by reading some poems. Hit Parade ratings of popular songs are obtained by questioning a sample of people, or finding out how many records have been sold in selected record shops, not in all the record shops.

Sampling is of great importance to manufacturers. They must be constantly on the alert to ensure that their products are of a high standard. In the production of some goods, every article can be tested e.g. every

T.V. set completed in a factory can be switched on to see that it is working. But how impractical it would be to test every ball bearing made in a factory to see that it was of standard size.

Many tests are by their nature destructive of the article e.g. testing the length of life of a light bulb or eating a cake to see if it is of good quality. If every single article was tested in this way, there would be nothing left to sell.

Manufacturers take samples of each batch of articles made to check the quality of the product.

In all the situations mentioned above, we assume that the sample gives us information about the whole collection or 'population'. We must recognise that whenever we obtain information from a sample and assume that this information holds good for the whole 'population', we run the risk of making a mistake.

In statistics we learn to pick out samples so that the risk of mistake is small. We reduce the risk by using more than one sample, by using samples of large size and by selecting **random samples.**

A random sample is one which is chosen in such a way that every possible member of the sample has an equal chance of being picked. For example, you have a number of even-sized marbles in a box—some red and some green; you give the box a good shake up and pick out five marbles without looking. You have picked out a random sample of five marbles, each marble in the box having had an equal chance of being chosen.

In a card game, the cards are shuffled and dealt out so that each player gets a random sample.

It may seem a simple matter to obtain a random sample, but in practical situations, particularly where human populations are concerned, it is far from simple; in fact, it is one of the most tricky jobs of a statistician (see Chapter 1).

Even in the card game mentioned above, what if the person shuffling and dealing the cards was a card sharp? The cards dealt to each player would certainly not be random samples—they would be distinctly biased.

Notation used for samples and populations

The **population** is the general term given to the aggregate of whatever we are considering. It only means a human population when people are being considered. A **sample** is a part of the population.

The following notation is used:

	Sample	Population
Number of variates	N	n
Mean	$\overline{X}$	μ
Standard deviation	s	σ

A numerical characteristic of a population such as its mean or standard deviation is called a **parameter.**

A quantity calculated from a sample such as its mean and standard deviation is called a **statistic.**

Since parameters of a given population are based upon *all* its variates, they are fixed for that population. On the other hand, since statistics are based upon only a *part* of the population, they usually vary from sample to sample.

How to choose a random sample

If the population we are considering is small, we can allot a numbered card or disc to each member (all the cards or discs being identical in size), and mixing the cards thoroughly, pick out the required sample.

Each member whose card was picked is in the sample.

This, however, is not practical where the number of the population is large. Here we can use tables of **random numbers.** These are four digit numbers selected from census reports and their randomness has been confirmed by numerous statistical tests.

This is a sample of twenty sets of random numbers.

2952	9792	7979	7002	8126
4167	2762	7203	5911	6111
2370	6107	3563	5356	3170
0560	9025	6008	1089	1300

Each member of the population is given a number. If you have decided on a sample of size 100, then 100 random numbers are picked and the members of the population with these numbers constitute the desired sample.

You can make up sets of random numbers for yourselves, each of the digits being obtained in two throws of an unbiased six-sided die, according to the following method.

The digits from 0–9 are divided into two equal groups (see third row of table).

The first throw of the die decides from which group the random digit will come.

If 1, 2 or 3 is scored, group 1 is chosen.
If 4, 5 or 6 is scored, group 2 is chosen.

On the second throw of the die any number, except 6, will determine the random digit. (We do not require the 6, since there are only five random digits in each group.) If a 6 should be scored, throw the die again until another number is obtained.

	Group 1					Group 2				
First throw	1, 2 or 3					4, 5 or 6				
Second throw	1	2	3	4	5	1	2	3	4	5
Random digit	0	1	2	3	4	5	6	7	8	9

Example

First Throw	Second Throw	Number Selected
2	4	3
5	4	8
4	(6) 1	5
3	(6) (6) 1	0
3	5	4

Thus the random number found is 38504.

Obviously with this method, we can choose Random Numbers with any desired number of digits.

The following experiment may be done by the class working individually or in pairs.
Equipment—1 die and shaker.

Determine 20 Random Numbers between 0 and 99. Calculate their mean and standard deviation. Compare the results for the mean and standard deviation of the whole class. (The mean should be approx. 50 and standard deviation approx. 30.)

Practical work

Experimental laboratory work can be conducted in order to test the validity of sampling.

EXPERIMENT *To estimate the percentage of red discs in a population of 400 discs by sampling*

Four hundred cardboard discs were used is this experiment, 350 of them being white and the other 50 red, i.e. $12\frac{1}{2}\%$ of the discs were red. Apart from colour, the discs were identical.

The discs were placed in a fairly large box and well shaken up. Random samples of varying sizes were drawn from the box, the discs being replaced after each draw. The percentage of red discs obtained in each sample was calculated as shown below.

Results

Sample—size 200 (50 % sample).
1. Number of red discs = 23
2. ,, ,, ,, ,, = 23
3. ,, ,, ,, ,, = 19
4. ,, ,, ,, ,, = 19
5. ,, ,, ,, ,, = 21

Percentage of red discs = $11\frac{1}{2}\%$
,, ,, ,, ,, = $11\frac{1}{2}\%$
,, ,, ,, ,, = $9\frac{1}{2}\%$
,, ,, ,, ,, = $9\frac{1}{2}\%$
,, ,, ,, ,, = $10\frac{1}{2}\%$
Mean percentage of red discs = $10\frac{1}{2}\%$.

Sample—size 120 (30 % sample)
1. Number of red discs = 11
2. ,, ,, ,, ,, = 15
3. ,, ,, ,, ,, = 12
4. ,, ,, ,, ,, = 13
5. ,, ,, ,, ,, = 12

Percentage of red discs = $9\frac{1}{6}\%$
,, ,, ,, ,, = $12\frac{1}{2}\%$
,, ,, ,, ,, = 10%
,, ,, ,, ,, = $10\frac{5}{6}\%$
,, ,, ,, ,, = 10%
Mean percentage of red discs = $10\frac{1}{2}\%$.

Sample—size 100 (25 % sample)
1. Number of red discs = 10
2. ,, ,, ,, ,, = 10
3. ,, ,, ,, ,, = 18
4. ,, ,, ,, ,, = 13
5. ,, ,, ,, ,, = 14

Percentage of red discs = 10%
,, ,, ,, ,, = 10%
,, ,, ,, ,, = 18%
,, ,, ,, ,, = 13%
,, ,, ,, ,, = 14%
Mean percentage of red discs = 13%.

Sample—size 40 (10 % sample)
1. Number of red discs = 3
2. ,, ,, ,, ,, = 2
3. ,, ,, ,, ,, = 3
4. ,, ,, ,, ,, = 4
5. ,, ,, ,, ,, = 6

Percentage of red discs $= 7\frac{1}{2}\%$
,, ,, ,, ,, $= 5\%$
,, ,, ,, ,, $= 7\frac{1}{2}\%$
,, ,, ,, ,, $= 10\%$
,, ,, ,, ,, $= 15\%$
Mean percentage of red discs $= 9\%$.

Sample—size 20 (5% sample).
1. Number of red discs $= 4$
2. ,, ,, ,, ,, $= 3$
3. ,, ,, ,, ,, $= 4$
4. ,, ,, ,, ,, $= 3$
5. ,, ,, ,, ,, $= 3$

Percentage of red discs $= 20\%$
,, ,, ,, ,, $= 15\%$
,, ,, ,, ,, $= 20\%$
,, ,, ,, ,, $= 15\%$
,, ,, ,, ,, $= 15\%$
Mean percentage of red discs $= 17\%$.

As seen from the results, the percentages of red discs in the individual samples vary quite a bit—more so in the smaller samples, but the mean of the samples does not show such a big variation.

Let us compare the mean percentages of red discs for the different sample sizes with the actual percentage of the population.

Percentage of red discs in the population	$= 12\frac{1}{2}\%$
Mean percentage of red discs of sample size	$200 = 10\frac{1}{2}\%$
,,	$120 = 10\frac{1}{2}\%$
,,	$100 = 13\%$
,,	$40 = 9\%$
,,	$20 = 17\%$

From these results, we can see that the three larger sample sizes give an answer fairly close to the actual percentage of red discs in the population. The two smaller sample sizes give much poorer answers.

From this experiment it would seem that a 50%, 30% or 25% sample gives comparable results (in fact the 25% sample gave the result closest to the correct answer), but the smaller sized samples gave fairly poor results.

Note: Where the population is *very large* a 10% sample gives good results.

It would also seem to be rather unwise to take a single sample and on that basis draw conclusions about the population.

In practice, then, a number of fairly large samples should be used and the mean of the samples calculated.

A number of experiments similar to that described above may be done. Try some out for yourselves, choosing various sizes of samples. (If you use multiples of ten for your population, samples, etc, it will make the calculations easier.)

Suggestions for experiments
1. Use identical counters or discs or small squares of carboard with a percentage of them different in colour, or marked in some way.
2. Use marbles of two different colours (but the same size and weight.)
3. Use a sampling bottle.
   A sampling bottle is a small bottle with a clear tube attached to the top. Small round beads are put into the bottle, a number of them being coloured or marked in some way. The length of the

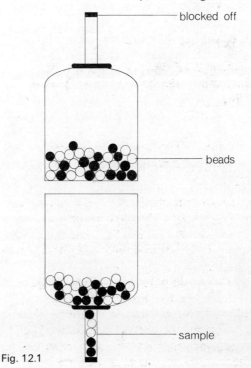

Fig. 12.1

tube is such that when the bottle is inverted only a certain number of beads can collect and be seen in the tube. The beads in the tube are the sample, and the coloured beads can be counted easily.

Sampling bottles can be bought or made quite easily with some help from the science department. Different lengths of tube can be used to give different sized samples. The small glass beads (obtained in different colours) used in science for density experiments etc. are ideal for the purpose.

Sampling does not have to be restricted to laboratory experiments. Try some sampling in your own class. Measure the heights of all the pupils in the class and calculate the mean height. Now assign a number to each pupil and have cards or discs with these numbers on them. Shake up the cards and draw out a 25% sample. Pick out the pupils with the numbers in the sample and calculate their mean height. Replacing the cards each time, draw several samples of this size calculating the mean heights for each sample. Compare these mean heights with the mean height of the whole class.

Now, calculate the mean of the mean heights of the samples and see how this compares with the class mean.

Repeat this procedure using a different size of sample.

What have you found out about the mean heights obtained from the samples compared to the class mean?

Sampling distributions

The results of an arithmetic test for speed and accuracy given to all first year pupils (179) in a secondary school are shown below.

Mean numbers of sums correct = 4.14.

We wished to use this distribution for sampling purposes and decided the easiest way to do this was to have 179 identical cards marked according to the number of correct sums,

i.e.   4 cards were marked 0
        10   ,,   1
        20   ,.   2
                          etc.

(Both sides of the cards were marked so that there would be no need to turn them over)

The cards were placed in a fairly large box and shaken up well. A sample of 40 cards was drawn out and the mean calculated, as shown below. The cards were replaced and the procedure repeated until 50 samples in all had been taken (all samples of size 40). In each case the mean of the sample was calculated.

Sample 1

No. of Sums correct (X)	Freq. (f)	fX
0	0	0
1	2	2
2	6	12
3	8	24
4	6	24
5	7	35
6	7	42
7	1	7
8	2	16
9	1	9
10	0	0
	$\sum f = 40$	$\sum fX = 171$

$$\overline{X} = \frac{\sum fX}{f}$$

$$= \frac{171}{40}$$

$$= 4.275$$

approx. 4.28

Mean of Sample 1 = 4.28

Number of sums correct	0	1	2	3	4	5	6	7	8	9	10
Frequency	4	10	20	33	39	29	25	13	4	2	0

124

The means of the samples (or sample means) were formed into a frequency distribution—a **sampling distribution** as shown below.

Sampling distribution

Sample Means	Frequency
3.50–3.59	1
3.60–3.69	1
3.70–3.79	3
3.80–3.89	3
3.90–3.99	5
4.00–4.09	5
4.10–4.19	6
4.20–4.29	8
4.30–4.39	10
4.40–4.49	2
4.50–4.59	4
4.60–4.69	2

The mean of this sampling distribution was calculated and found to be 4.17 i.e. the mean of the sample means was 4.17.

Distribution of the sample means

A variable X has a normal distribution with mean μ and standard deviation σ.

If observations are collected, not individually but as random samples of size N, then the means of the samples can be formed into a frequency distribution which will be approximately normal. This distribution is called a sampling distribution, the variable this time being $\bar{X}$.

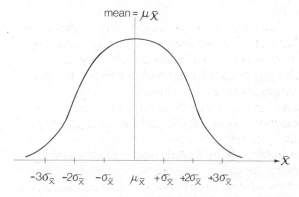

Fig. 12.3

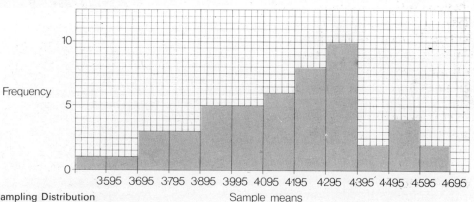

Fig. 12.2 Sampling Distribution

This histogram shows that the sampling distribution is reasonably near to a normal distribution considering that the number of samples is only 50.

The mean of the sampling distribution (or mean of the sample means) is 4.17, very close to the mean of the population 4.14.

Try an experiment like this yourselves with some data of your own. You should use a distribution which is approximately normal with a fairly large total frequency.

The mean of the sampling distribution is called $\mu_{\bar{x}}$ to distinguish it from the population mean μ. Similarly the standard deviation of the sampling distribution is called $\sigma_{\bar{x}}$.

There is a definite correlation between the mean and standard deviation of the sampling distribution, and the mean and standard deviation of the population.

For practical purposes we can assume that the mean of the sampling distribution (i.e.

the mean of the sample means) $\mu_{\bar{x}}$ is approximately the same as the population mean.

$$\text{i.e. } \mu_{\bar{x}} = \mu$$

Also, $\sigma_{\bar{x}}$ the standard deviation of the sampling distribution is equal to the population standard deviation, σ, divided by the square root of N, the number in the samples.

$$\text{i.e. } \sigma_{\bar{x}} = \frac{\sigma}{\sqrt{N}}$$

This standard deviation, $\sigma_{\bar{x}}$, of the sample mean $\bar{X}$ from the population mean μ is known as the **standard error** of the sample mean from the population mean.

$$\text{i.e. The Standard Error} = \frac{\sigma}{\sqrt{N}}$$

Obviously as N, the size of the sample, increases, the standard error of the sample mean from the population mean i.e. the difference between the sample mean and the population mean, becomes smaller and the standard deviation, s, of a single *large* sample approximates to, σ the standard deviation of the population i.e. s = σ.

We standardise a score X of a normal distribution with mean μ and standard deviation σ by using this formula.

$$\text{Standard score, } Z = \frac{X - \mu}{\sigma}$$

In the same way we may standardise a score $\bar{X}$ of the sampling distribution with mean $\mu_{\bar{x}}$ and standard deviation $\sigma_{\bar{x}}$ by using the same formula in this way:

$$\text{Standard score, } Z = \frac{X - \mu_{\bar{x}}}{\sigma_{\bar{x}}}$$

Example 1
The mean height of 1000 university students is 171.25 cm and the standard deviation is 6.25 cm. Find the probability that in a sample of 100 students, the mean will be greater than 172.50 cm.

Solution

$$\text{Population}: \mu = 171.25 \text{ cm}$$
$$\sigma = 6.25 \text{ cm}$$
$$\text{Sampling}: \mu_{\bar{x}} = \mu = 171.25 \text{ cm}$$

$$\sigma_{\bar{x}} = \frac{\sigma}{\sqrt{N}}$$
$$= \frac{6.25}{\sqrt{100}}$$
$$= \frac{6.25}{10}$$
$$= 0.625 \text{ cm}$$

We must find the area under the normal curve to the right of $\bar{X} = 172.50$ (the mean of the sample of 100 students).

$$Z = \frac{\bar{X} - \mu_{\bar{x}}}{\sigma_{\bar{x}}}$$

(We are standardising the sample mean $\bar{X} = 69$ in the sampling distribution)

$$= \frac{172.50 - 171.25}{0.625}$$
$$= \frac{1.25}{0.625}$$
$$= 2$$

Fig. 12.4

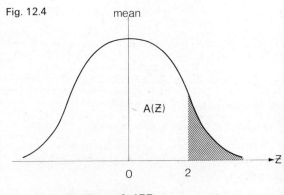

$$\therefore A(Z) = 0.477$$
$$\therefore \text{Reqd. area} = 0.5 - 0.477$$
$$= 0.023$$

i.e. The probability of the sample mean being greater than 172.50 cm = 0.023

Example 2

Assume that the heights of 3000 men are normally distributed with mean 172.5 cm and standard deviation 7.5 cm. If 50 samples of 25 men each (assuming replacement) are obtained, what would be the expected mean and standard deviation of the resulting sampling distribution? In how many samples would you expect to find the mean between 168.75 and 173.75 cm?

Solution

$$\text{Population}: \mu = 172.5 \text{ cm}$$

$$\sigma = 7.5 \text{ cm}$$

$$\text{Sampling}: \mu_{\bar{x}} = \mu = 172.5 \text{ cm}$$

$$\sigma_{\bar{x}} = \frac{\sigma}{\sqrt{N}}$$

$$= \frac{7.5}{\sqrt{25}}$$

$$= \frac{7.5}{5}$$

$$= 1.5 \text{ cm}$$

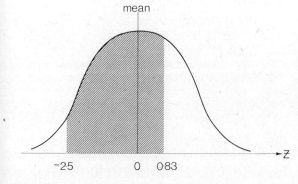

Fig. 12.5

We must find the area under the normal curve

$$\text{from } \bar{X}_1 = 168.75 \text{ to } \bar{X}_2 = 173.75$$

$$Z_1 = \frac{\bar{X}_1 - \mu_{\bar{x}}}{\sigma_{\bar{x}}}$$

$$= \frac{168.75 - 172.50}{1.5}$$

$$= \frac{-3.75}{1.5}$$

$$= -2.5$$

$$Z_2 = \frac{\bar{X}_2 - \mu_{\bar{x}}}{\sigma_{\bar{x}}}$$

$$= \frac{173.75 - 172.50}{1.5}$$

$$= \frac{1.25}{1.5}$$

$$= 0.83$$

$$A(Z_1) = 0.494$$

$$A(Z_2) = 0.297$$

$$\text{Reqd. area} = 0.494 + 0.297$$

$$= 0.791$$

$\therefore$ Probability of the sample mean being between 168.75 and 173.75 cm = 0.791
Expected number of samples

$$= \text{prob.} \times \text{no. of samples}$$

$$= 0.791 \times 50$$

$$= 39.55$$

$$= \text{approx. } 40$$

EXERCISE A

1. In a normal distribution with mean 73.5 and standard deviation 2.80, find the probability that in a sample of 100 variates, the mean will be
   a) more than 74
   b) less than 72.6

2. In a normal distribution with mean 53.6 and standard deviation 3.50, find the probability that in a sample of 36 variates, the mean will be
   a) less than 52.8
   b) between 53.2 and 54.

3. In a savings bank, the average account is £160.85 with a standard deviation of £18. What is the probability that a group of 81 accounts taken at random, will show an average deposit of £165 or more?

127

4. The heights of a certain group of adults are normally distributed with a mean of 170.75 cm and a standard deviation of 6.25 cm. If 25 people are chosen at random from the group, what is the probability that their mean height will be 172.25 cm or more?

5. If all possible samples of size 36 are drawn from a normally distributed population with mean 30 and standard deviation 3, within what range will the middle 50% of the sample means lie?

6. The weekly wages of a certain industry are normally distributed with a mean of £20. If 10% of the mean wages of samples of 25 workers fall below £19.50, what is the standard deviation of weekly wages in this industry?

7. (a) Assuming that the heights of men are normally distributed with a standard deviation of 5 cm, how large a sample should be taken in order to be fairly sure (with a probability of 0.95) that the sample mean does not differ from the population mean by more than 1.0, in absolute value?

(b) How large a sample would need to be taken to be 99% sure?

8. A normal population has a standard deviation of 3. How large a sample should be taken in order to be 99% sure that the sample mean differs from the true mean by less than 0.5?

Confidence limits

Suppose we have a sample of N variates with mean $\bar{x}$ and standard deviation s selected from a normal population. Let us consider the problem of estimating the mean of the population from this single sample. It is obviously impossible to determine the population mean precisely, since this sample could have been taken from any population with roughly the same size of mean. What we can do, is establish **limits** within which the mean of the population will fall with a specified probability or **confidence.**

128

To illustrate this, suppose a random sample of 100 variates is taken from a normal population. The mean of the sample is found to be 30 and the standard deviation is 5. From this one sample we will try to estimate the mean of the population with a probability of 95% (or 95% confidence).

With a probability of 0.95 the population mean will fall in the shaded area of the diagram.

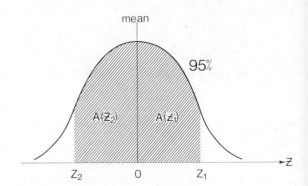

Fig. 12.6

i.e. $A(Z_1) = 0.475$

$A(Z_2) = 0.475$

$\mu_{\bar{x}} = \mu$

$\sigma_{\bar{x}} = \dfrac{\sigma}{\sqrt{N}}$

We do not know either $\sigma_{\bar{x}}$ or σ. However, the standard error, $\dfrac{\sigma}{\sqrt{N}}$, of the sample mean from the population mean i.e. the difference between the sample mean and the population mean becomes smaller as N becomes larger, so that for a **large sample,** we may assume that the standard deviation of the sample, s, approximates to σ.

Thus $\sigma_{\bar{x}} = \dfrac{\sigma}{\sqrt{N}}$

$= \dfrac{s}{\sqrt{N}} = \dfrac{5}{\sqrt{100}} = \dfrac{5}{10} = 0.5$

From the tables we find that

$$Z = +1.96$$
$$Z = -1.96$$

i.e. we must say

$$Z = \pm 1.96$$

$$Z = \frac{\bar{X} - \mu_{\bar{x}}}{\sigma_{\bar{x}}}$$

$$\pm 1.96 = \frac{30 - \mu}{0.5}$$

$$\therefore \pm 1.96 \times 0.5 = 30 - \mu$$

$$\therefore \mu = 30 - (\pm 1.96 \times 0.5)$$
$$= 30 \pm 1.96 \times 0.5$$
$$= 30 \pm 0.980$$

With 95% confidence we can say that the population mean lies within the limits 30 ± 0.980.

(The answer is very often left in this form)

We could also have given the answer like this:

The 95% confidence limits for the population mean are 30 ± 0.980.

(These limits define the interval which would contain the population mean with a probability of 95%).

The interval between the confidence limits is called the confidence interval.

Fig. 12.7

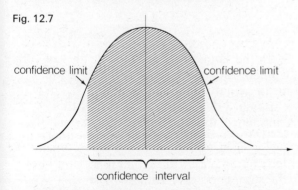

With 95% confidence limits, we find Z_1 and Z_2 by using the tables to find the Z score corresponding to these areas

$$A(Z_1) = \tfrac{1}{2} \text{ of } 0.95 = 0.475$$
$$A(Z_2) = \tfrac{1}{2} \text{ of } 0.95 = 0.475$$

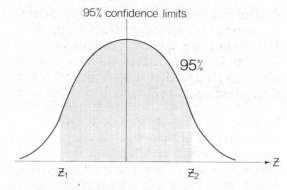

Fig. 12.8 (a) 95% Confidence Limits

With 99% confidence limits, we find Z_1 and Z_2 by using these areas.

$$A(Z_1) = \tfrac{1}{2} \text{ of } 0.99 = 0.495$$
$$A(Z_2) = \tfrac{1}{2} \text{ of } 0.99 = 0.495$$

Similarly we can find Z for any specified confidence.

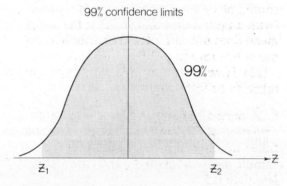

Fig. 12.8 (b) 99% Confidence Limits

Note: We can only use the above method for estimating a population mean when the sample is large, where it is permissable to assume that the standard deviation of the sample can be taken as an approximation of the standard deviation of the population.

Example

A sample of 80 variates has a mean of 60 with a standard deviation of 5.5. Find the 98% confidence limits for the mean of the population.

129

Solution

Here, $\bar{X} = 60$

$s = 5.5$

$N = 80$

And, $\mu_{\bar{x}} = \mu$

$$\sigma_{\bar{x}} = \frac{\sigma}{\sqrt{N}}$$

$$= \frac{s}{\sqrt{N}} \quad \text{(Since the sample is large } \sigma \text{ can be taken as s)}$$

$$= \frac{5.5}{\sqrt{80}} = \frac{5.5}{8.94} = 0.62$$

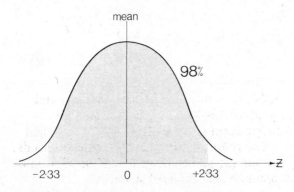

Fig. 12.9

With a probability of 0.98

$$A(Z_1) = \tfrac{1}{2} \text{ of } 0.98 = 0.49$$

and $A(Z_2) = 0.49$

$$\therefore Z_1 = +2.33$$

$$\therefore Z_2 = -2.33$$

i.e. $Z = \pm 2.33$

$$Z = \frac{\bar{X} - \mu_{\bar{x}}}{\sigma_{\bar{x}}}$$

$$\therefore \pm 2.33 = \frac{60 - \mu}{0.62}$$

$$\therefore \pm 2.33 \times 0.62 = 60 - \mu$$

$$\therefore \mu = 60 \pm 2.33 \times 0.62$$

$$= 60 \pm 1.4446$$

$$= 60 \pm 1.44$$

130

i.e. The 98% confidence limits for the population mean are

$$60 \pm 1.44$$

EXERCISE B

1. In a sample of 60 variates, the mean is 30 and the standard deviation 3.8. Find the limits within which the population mean is expected to lie, with a probability of 95%.

2. A sample of 100 variates has a mean of 65 and a standard deviation of 2.5. Find the 99% confidence limits for the mean of the population.

3. Find the 98% confidence limits for the mean of a population, if a sample of 90 variates taken from the population has a mean of 12.6 and a standard deviation of 3.

4. Suppose the standard deviation of heights in males is 6.25 cm. Two hundred male students in a large university are measured and their average height is found to be 171.05 cm. Within what range of heights would you expect to find the mean height of the men in this university with a confidence of 95%?

5. In the same university, the heights of 100 women students were measured and the mean height was found to be 161.25 cm with a standard deviation of 5.5 cm. With 95% confidence, find the limits within which the mean height of the women in that university lie.

6. A sample of 50 light bulbs was taken from a very large batch produced in a factory. The mean life of the bulbs was found to be 2.5 years with a standard deviation of 6 months. Estimate the mean life of the batch of bulbs with a probability of 99%.

13 SIGNIFICANCE TESTS

In statistics, we often wish to find out if something that occurs is within the normal range of events or whether it is so far out with the normal range that something is wrong somewhere.

For instance, we find that in tossing a coin 100 times, we obtain 60 heads. Is this a 'reasonable' number of heads to get or is there something wrong with the coin i.e. is it biased in some way? (We assume the coin is being tossed properly.)

This is the kind of thing we are concerned with in significance testing. What we have to do is to formulate a **statistical hypothesis** and then test this hypothesis.

Suppose we wish to test the honesty of the coin mentioned above. We first make the assumption that the coin is honest i.e. that it is equally likely to show a head or a tail. This is our statistical hypothesis (or **null hypothesis**). Then working on this basis, we calculate the probability of obtaining 60 or more heads out of 100 tosses. (We do not use exactly 60 heads since the probability of obtaining exactly 60 heads is extremely small).

Method

$$\text{Hypothesis: } p = \tfrac{1}{2} \text{ (head)}$$

$$q = \tfrac{1}{2}$$

$$N = 100$$

$$\mu = Np$$

$$= 100 \times \tfrac{1}{2}$$

$$= 50$$

$$\sigma = \sqrt{Npq}$$

$$= \sqrt{\frac{100}{1} \times \tfrac{1}{2} \times \tfrac{1}{2}}$$

$$= 5.$$

We have now calculated the mean and standard deviation of the binomial distribution given by $(p+q)^{100}$.

The score we are interested in is 60 heads. To include 60 heads we take X as 59.5; and calculate the standard score.

$$Z = \frac{X-\mu}{\sigma}$$

$$= \frac{59.5-50}{5}$$

$$= \frac{9.5}{5}$$

$$= 1.9$$

$$A(Z) = 0.471$$

$$\therefore \text{ Required area} = 0.5 - 0.471$$

$$= 0.029$$

i.e. the probability of obtaining 60 or more heads = 0.029

We have calculated that if an honest coin is tossed 100 times, the probability of obtaining 60 or more heads is 0.029 – certainly not a high probability.

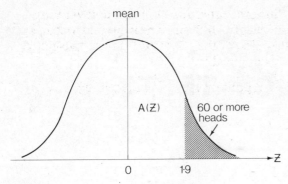

Fig. 13.1

From this result we are justified in drawing one of two conclusions:

1. The hypothesis is correct but a rare event has occurred.
2. The hypothesis is not correct.

In statistics, we decide that if the probability is less than a given value α, called the **significance level,** then the hypothesis is not correct; if the probability is greater than α, then the hypothesis is correct.

We leave ourselves open to two types of error here.

A Type I error is when a correct hypothesis is rejected.

A Type II error is when an incorrect hypothesis is accepted.

If the calculated probability is less than α, indicating the hypothesis is false, the result is said to be **significant.**

The value given to α may be different in different situations. So we must always state that our conclusions are based on a certain level of significance. We customarily take α to have the value 0.05 (the 5% level of significance).

Looking back at our example about the coin.

The probability of 60 or more heads $= 0.029$

Significance level, $\alpha = 0.05$

The probability is less than α.

This is a *significant result.*

Thus, at the 5% level of significance the hypothesis is rejected i.e. the coin is not honest.

132

If we were testing a new and potentially dangerous drug, a much smaller level of significance would be chosen, perhaps $\alpha = 0.01$.

One- and two-tailed tests

In the significance test on the coin in the last section, we used only one end or 'tail' of the normal curve. This was a **one-tailed test.** If both 'tails' are used, we refer to this as a **two-tailed test.**

The decision to use a one- or two-tailed test must be reached by careful consideration of the question to be answered. If the interest in the problem is restricted to the fact that a very low or very high result is obtained (but not both) a one-tailed test is called for—otherwise a two-tailed test should be applied. In case of doubt, a two-tailed test is recommended.

When $\alpha = 0.05$, a result in a two-tailed test is significant if Z falls in the shaded part of figure 13.2 (a) i.e. if Z is greater than $+1.96$ or less than -1.96, where 1.96 is the value of Z obtained from the tables for each of the shaded areas, equal to 0.025 (half of 0.05).

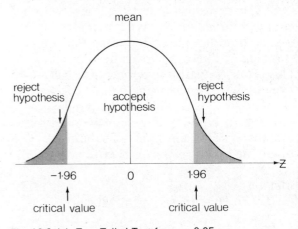

Fig. 13.2 (a) Two Tailed Test for $\alpha = 0.05$

Similarly in a one-tailed test, the result is significant if Z falls in the shaded area of figure 13.2 (b) i.e. if Z is greater than 1.64, where 1.64 is the Z value for the shaded area of 0.05.

One tailed test for $\alpha = 0.05$

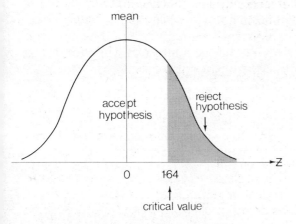

Fig. 13.2 (b) One Tailed Test for $\alpha = 0.05$

These values of Z (± 1.96 and 1.64) are called **critical values.**

Example
According to genetic theory, we know that certain crosses of peas should give tall plants and short plants in a ratio of 4:1. In a particular experiment, 170 tall plants and 30 short plants were obtained. Is this a significant deviation from the theory on the basis of the 5 % level of significance?

Solution
Here we are testing the hypothesis that the probability of obtaining tall plants is $\frac{4}{5}$.

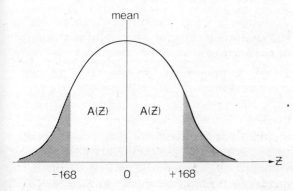

Fig. 13.3

Since this question concerns a deviation from theory, and this deviation may occur on either side of the expected result, a two-tailed test is applicable.

$$\text{Hypothesis: } p = \frac{4}{5} \text{ (tall plants)}$$

$$q = \frac{1}{5}$$

$$N = 200$$

$$\mu = Np$$

$$= 200 \times \frac{4}{5}$$

$$= 160$$

$$\sigma = \sqrt{Npq}$$

$$= \sqrt{\frac{200}{1} \times \frac{4}{5} \times \frac{1}{5}}$$

$$= \sqrt{32}$$

$$= 5.66$$

The score we are interested in is 170 tall plants.
So, we take x as 169.5 to include 170

$$Z = \frac{x - \mu}{\sigma}$$

$$= \frac{169.5 - 160}{5.66}$$

$$= \frac{9.5}{5.66}$$

$$= 1.68$$

$$A(Z) = 0.454$$

Since a two-tailed test is used
Required area for $Z > 1.68 = 0.5 - 0.454$

$$= 0.046$$

Required area for $Z < -1.68 = 0.046$

$$\text{Total area} = 0.092$$

$$\text{i.e. probability} = 0.092$$

$$\alpha = 0.05$$

The probability is greater than α.

On the 5% level of significance, this result is not significant, i.e. there is no significant deviation from the genetic theory.

Practical work

In a well known advertisement for margarine, we are told that 8 out of 10 people cannot tell Stork from butter. Test this claim by the manufacturer. Discuss in class, what your hypothesis should be, and how many people you should test.

What precautions will you have to take in carrying out this experiment?

You might repeat the experiment using a different type of butter.

Many people say that they can always tell Stork from butter. Find one or more of these people and test their claim.

What will your hypothesis be this time?

EXERCISE A

Unless told otherwise, use the 5% level of significance.

Answer each question exactly in the form in which it is asked.

1. A die is rolled 120 times. If a 6 is obtained 26 times, have we cause to doubt the honesty of the die?

2. A coin is tossed 400 times. Heads turn up 250 times. Is this an honest coin?

3. A pair of dice are rolled 360 times. Two sixes show 15 times. Have we any reason to think the dice are 'loaded'?

4. From experience it is known that 30% of a certain kind of seed germinate. If in an experiment only 70 out of 300 seeds germinate, is this a significantly poor germination at the 1% level of significance?

5. In a factory manufacturing a certain product it is known from long experience that 7% of the articles do not come up to standard and have to be discarded. A new worker who has been taken on has made 400 articles of which 36 are defective. Is there any reason to doubt the man's ability to do the job?

6. According to genetic theory, the offspring of a certain cross between rabbits should be white to not white in a ratio of 3:7. In one experiment the number of rabbits not white was 71, and the number of white rabbits was 19. Was this result consistent with the genetic theory?

7. A sample of 100 has a mean of 55 and a standard deviation of 9.
   a) Does the mean of this sample differ significantly at the 5% level of significance from a population mean of 56?
   b) Is the sample mean significantly better at the 5% level than a population mean of 53 (using a one-tailed test)?
   c) Calculate from the sample the 95% confidence limits for the population mean.

8. The heights of adult men in a certain town have a mean of 171.55 cm with a standard deviation of 6.0 cm. A sample of 144 men living in a slum district is found to have a mean height of 170.00 cm. Does this indicate that the residents of the slums are significantly retarded in growth on the basis of the 1% level of significance?

9. A manufacturer of springs has established from several years experience that the springs he makes have a mean stretching point of 14.6 kg (i.e. the springs do not go back into shape after this point) with a standard deviation of 2.2 kg. After a change in the manufacturing process, a sample of 50 springs is taken and the mean stretching point is found to be 13 kg with a standard deviation of 2 kg. Has the new process had a significantly damaging effect on the strength of the springs?

10. A car manufacturer claims that his cars use an average of 2.80 gallons of petrol for each 100 miles. A car salesman tests 40 cars made by this company, for petrol mileage, and he finds the average petrol consumption to be 2.88 gallons for each 100 miles with a standard deviation of 0.2 gallons. Do these results cast doubts on the manufacturer's claim (1% level)?

ANSWERS

Chapter 2
Exercise A   Page 16
1. *a)*   5.63   *b)*   10.1   *c)*   10   *d)*   0.0068   *e)*   6   *f)*   120
2. *a)*   11.7   *b)*   9.8   *c)*   69.0   *d)*   0.1   *e)*   7.0   *f)*   15.8
   g)   15.8 or 15.9   *h)*   15.9
3. *a)*   600,000   *b)*   580,000   *c)*   583,000   *d)*   582,700   *e)*   582,730

Exercise B   Page 17

	a)	*b)*	*c)*
1.	1 cm	0.5 cm	7.5 cm, 6.5 cm
2.	1 m	0.5 m	186.5 m, 185.5 m
3.	1 h	0.5 h	17.5 h, 16.5 h
4.	0.1 kg	0.05 kg	9.25 kg, 9.15 kg
5.	0.1 g	0.05 g	12.85 g, 12.75 g
6.	0.001 litres	0.0005 l	2.6835 l, 2.6825 l
7.	0.01 mile	0.005 mile	1.215 mile, 1.205 mile
8.	0.1 s	0.05 s	16.55 s, 16.45 s

Exercise C   Page 18

	a)	*b)*		*a)*	*b)*
1.	0.5 m	$\dfrac{1}{300}$	6.	$\dfrac{1}{70}$	1.4%
2.	0.5 kg	$\dfrac{1}{70}$	7.	$\dfrac{1}{34}$	2.9%
3.	0.05 l	$\dfrac{1}{30}$	8.	$\dfrac{1}{500}$	0.20%
4.	0.05 s	$\dfrac{1}{46}$	9.	$\dfrac{1}{2500}$	0.040%
5.	$\dfrac{1}{14}$	7.1%	10.	$\dfrac{1}{30\,000}$	0.0033%

Exercise D   Page 18
1. *a)*   14.80 cm,   14.60 cm   *b)*   33.0 g,   31.0 g
   c)   41.0 m,   39.0 m   *d)*   40.0 litres,   38.0 litres
2. *a)*   1.0 m   *b)*   1.0 kg   *c)*   0.10 g   *d)*   0.010 cm

3. *a)* 15.0 cm, 13.0 cm   *b)* 6.0 kg, 4.0 kg
   c) 7.40 m, 7.20 m   *d)* 3.70 litres, 3.50 litres
4. *a)* 1.0 m   *b)* 1.0 ml   *c)* 0.55 cm   *d)* 0.55 g

Exercise E   Page 19

1. 503.75	2. 19.65	3. 34.77	4. 165.1425
5. 10.47	6. 9.29	7. 14.75	8. 0.43
9. 0.64	10. 8.11		

Chapter 3

Exercise A   Page 20

1. Continuous	2. Discrete	3. Continuous	4. Discrete
5. Continuous	6. Continuous	7. Discrete	8. Discrete
9. Continuous	10. Continuous	11. Continuous	12. Discrete

Exercise B   Page 22

1.

Heights	142	144	146	148	150	152	154	156	158	160	162	164
Frequency	1	1	0	3	5	8	10	4	6	1	0	1

  a) Continuous   *b)* 22 cm   *c)* 154 cm   *d)* 0.25 or 25%
  e) 5   *f)* 5%   *g)* 67.5%

3.

No. of children	1	2	3	4	5	6	7	8	9	10
Frequency	2	4	5	2	4	7	1	1	1	2

  a) Discrete   *b)* 6 children   *c)* Families with no. children
  d) 5   *e)* 37.9%

5. *a)*

Goals	0	1	2	3	4	5
Frequency	7	11	10	4	3	1

  b) Discrete   *c)* 5 goals   *d)* 1 goal   *e)* 0.306 or 30.6%
  f) 22.2%

Exercise C   Page 23

1. *a)* 22   *b)* 13   *c)* 16   *d)* 0.18   *e)* 6   *f)* 6
2. *a)* The half sizes do not have as large frequencies as might be expected. We discovered that this trend was markedly obvious all through the school for all age groups of boys and girls. To try to explain this, some pupils visited shoe shops in the area and interviewed the managers. The two main explanations for this trend were that mothers were inclined to buy the next whole size of shoe to allow for growth, rather than take the half size; and that the cheaper make of shoes are often only made in whole sizes.
  b) 68   *c)* 5   *d)* 0.29 or 29%   *e)* 0.103 or 10.3%

4.

No. of Words	1	2	3	4	5	6	7	8	9	10	11	12	13
Frequency	1	0	2	0	2	4	3	11	7	3	3	1	1

  a) 38   *b)* 8   *c)* 12   *d)* 0.18   *e)* 8
  f) 3   *g)* 18
6. *a)* 32   *b)* 4   *c)* 8   *d)* 21.9%

Exercise D   Page 25

1.   *a)* 30   *b)* 8   *c)* 76   4. *a)* 3, 6

Exercise E Page 27

1.

Mark	1–5	6–10	11–15	16–20	21–25	26–30	31–35	36–40	41–45	46–50
Frequency	2	6	8	10	6	3	2	1	1	1

a) 23   b) 25.5   c) 5.5   d) 4th class   e) 0.2
f) 20%   g) 60%

3.

Rate	21–22	23–24	25–26	27–28	29–30	31–32	33–34	35–36	37–38
Frequency	2	7	10	8	9	5	1	1	1

a) 21.5   b) 26.5   c) 28.5   d) 3rd class   e) 0.2

4.

No. of rainy days	140–149	150–159	160–169	170–179	180–189	190–199	200–209
Frequency	1	4	4	7	11	10	9

No. of rain days	210–219	220–229	230–239	240–249	250–259	260–269
Frequency	2	5	4	3	2	3

a) 174.5   b) 189.5   c) 239.5   d) 5th class   e) 37
f) 56.9%

Exercise F Page 29

1. a) 65   b) 50.5   c) 125.5   d) 88   e) 2nd class
    f) 0.4   g) 20%   h) 7.7%
2. a) 80   b) 49.5   c) 44.5   d) 45–49 kg   e) 55–59 kg
    f) 52   g) 4th class   h) 0.3

Exercise H Page 34

1.

Heights (cm)	Cum. Freq.
146	1
146–148	1
–150	2
–152	5
–154	8
–156	12
–158	15
–160	20
–162	26
–164	27
–166	29
–168	30

a) 1
b) 60%
c) 3

2. a)

Time	Cum. Freq.
0– 20	0
– 40	2
– 60	6
– 80	13
–100	24
–120	52
–140	72
–160	88
–180	98
–200	104
–220	107

b) 6
c) 51.4%

3.

Weight (kg)	Cum. Freq.
35–39	1
35–44	3
35–49	6
35–54	14
35–59	22
35–64	26
35–69	27
35–74	27
35–79	28
35–84	29

a) 29
b) 6
c) 7

4. a)

Score	Freq.
65	1
66	2
67	2
68	8
69	7
70	7
71	5
72	2
73	2
74	2
75	2

b)

Score	Cum. Freq.
65	1
65–66	3
65–67	5
65–68	13
65–69	20
65–70	27
65–71	32
65–72	34
65–73	36
65–74	38
65–75	40

c) 20
d) 8

Chapter 4

Exercise A   Page 37

1.

	Mean	Median	Mode
a)	approx 9.2	10	11
b)	23	22.5	21
c)	3	3	3

2.   The mode, because it is the lowest.
3.   a) 162 cm   b) 158 cm or 162 cm   c) 162.1 cm
6.   The mode would be the easiest to find and the mean the hardest.
7.   Mean—£6.80   Median £6.20

Exercise B   Page 38

1.   Approx. 1.25 goals
3.   Approx. 3.9 children
5.   Approx. 9.3 words per line

Exercise C   Page 40

1.   Approx. 39.4 years
2.   Approx. 56.5 marks
3.   Approx. 39.7 years
4.   Approx. 77.4 hours

Exercise D   Page 42

1.   100
2.   Approx. 146.4 cm
3.   Approx. 116.08 g
4.   125.5 ships
5.   160.2 cm
6.   22.48 cm

Exercise E   Page 44

1. *a*) 29.5   *b*) 23.5   2. *a*) 23.8 *b*) 55%   *c*) 45%
3. 158.1 cm   4. 69.7 strokes   5. 55 kg
6. 122.5 secs   7. 572 deaths.

Chapter 5

Exercise A   Page 49

1. $13\frac{1}{3}$, 10, 9, 10, $11\frac{2}{3}$, 16, $16\frac{2}{3}$, $16\frac{2}{3}$.
2. 12.4, 10.0, 10.0, 14.0, 14.0, 14.6.
3. 23, $25\frac{1}{2}$, $25\frac{3}{4}$, 26, $28\frac{1}{4}$, $26\frac{1}{4}$, 28, $31\frac{1}{4}$, 32.
4. $193\frac{1}{2}$, $194\frac{1}{4}$, $189\frac{1}{2}$, $182\frac{3}{4}$, 178, $178\frac{1}{2}$, $183\frac{3}{4}$, $186\frac{1}{2}$, 181, 188, 194, $198\frac{3}{4}$, $211\frac{1}{4}$.
5. 204.3, 202.1, 199.6, 195.4, 194.8, 194.2, 196.1.
6. *a*) 1155, 1169, 1222, 1294, 1310, 1343, 1364, 1417, 1659.
   b) 1155, 1172, 1227, 1310, 1327, 1370, 1377, 1424, 1440.
   c) 1956, 2000, 2102, 2242, 2311, 2376, 2476, 2624, 2704.
   d) 1959, 2010, 2106, 2250, 2314, 2390, 2479, 2628, 2711.
7. £67, £68, £69, £68, £70, £71, £73, £74, £76, £75, £78, £85, £92, £92, £93, £96, £98, £100, £107, £107, £106, £104, £102, £96, £88.
8. 3.3%, 3.0%, 2.9%, 3.3%, 3.6%, 3.2%, 2.6%, 2.7%.
9. 14.6%, 14.6%, 14.4%, 14.5%, 14.3%, 14.0%, 13.5%, 13.0%, 12.6%.

Exercise B   Page 51

1. 8   2. 17   3. 50%   4. 7.5 gm per cm^3
5. 35.5   6. 63.6

Exercise C   Page 51

1. A 105   B 120   C 80   D 154   E 184
2. A £10.50 B £9.36   C £1.05   D £335   E £355
3. A 120   B 120   C 90   D 80   E 109
4. 147   5. 99

Exercise D   Page 53

1. 156   2. 119   3. 111   4. 114   5. 110

Chapter 6

Exercise A   Page 57

1.

	Range	Semi-interquartile Range
a)	6	1.5
b)	10	3
c)	10	3
d)	8	3

2. *a*) Median = 70.25 kg   *b*) Semi-interquartile range = 1.4 kg   *c*) 6 kg
   d) 27 boys   *c*) 22 boys.
3. *a*) 46 marks; 64 marks   *b*) 9.2 marks; 12.7 marks
   c) 76 pupils   *d*) 45.7%   *e*) 65 marks.
4. *a*) Boys: 162.2 cm   Girls: 161.5 cm   *b*) Boys: 2 cm   Girls: 1.25 cm
   c) 7 girls   *d*) 9 boys   *e*) 8 girls   *f*) 16 boys.
5.

	Median	Quartiles	Semi-interquartile Range
English	51	41, 59.5	9.25
Arithmetic	52.5	48, 57.5	4.75

1.   2.86   2. 4   3. 5.5

1. *a)* Mean = 7 ; S.D. = 3.12   *b)* Mean = 23.4; S.D. = 3.24
   c) Mean = 65 ; S.D. = 2.73   *d)* Mean = 35.5; S.D. = 2.96
   e) Mean = 29.2; S.D. = 7.60
2.   Mean = 77.7°F;   S.D. = 3.10°F   3. Mean = 15.5°F;   S.D. = 4.03°F
4.   Mean = 163.05 cm; S.D. = 6.37 cm   5. Mean = 156.47 cm; S.D. = 5.48 cm
7. North Scotland: mean = 19.9 days; S.D. = 1.93 days.
   East   Scotland: mean = 16.6 days; S.D. = 1.44 days.
   West Scotland: mean = 18.7 days; S.D. = 1.93 days.
8.   Mean = 24.5 marks; S.D. = 8.15
9.   Mean = 53.25 kg   ; S.D. = 9.65 kg
11. Mean = 9.76   ; S.D. = 1.71

	Mean	*Standard Deviation.*	
1.	152.3 cm	1.82 cm	
2.	53.2	1.32	
3.	£83.40	£13.30	
4.	24.0	9.05	
5.	46.6	13.5	(First Examination)
	58.3	21.2	(Second Examination)
6.	161.8 cm	4.12 cm	(Boys)
	161.6 cm	2.28 cm	(Girls)
7.	50.0	14.1	(English Examination)
	52.0	7.94	(Arithmetic Examination)

1.   May   2. *a)* 3rd exam   *b)* No—it was the same

3. *a)*

Pupil	*Mark*	*Deviate from mean*	*Standard Score*
Alan	68	68–64 = + 4	$+\frac{1}{2}$
Betty	72	72–64 = + 8	$+1$
Colin	56	56–64 = − 8	-1
David	60	60–64 = − 4	$-\frac{1}{2}$
Evelyn	80	80–64 = +16	$+2$
Fay	76	76–64 = +12	$+1\frac{1}{2}$

b)

Pupil	*Mark*	*Deviate from mean*	*Standard Score*
Alan	58	+ 9	$+1\frac{1}{2}$
Betty	52	+ 3	$+\frac{1}{2}$
Colin	43	− 6	-1
David	49	0	0
Evelyn	64	+15	$+2\frac{1}{2}$
Fay	55	+ 6	$+1$

c) Alan   *d)* Betty and Fay   *e)* Colin
4. Mary   5. Helen, Jean, Mary.

Chapter 7

1.   *a)* y = 2x−6   *b)* y = 7−x
      c) y = $\frac{1}{2}$x+3$\frac{1}{2}$ *d)* y = $\frac{2}{3}$x+1$\frac{2}{3}$

2.   a) $y = \frac{3}{2}x - \frac{1}{2}$ b) $y = 3\frac{1}{2} - \frac{1}{4}x$

3. $y = \frac{1}{2}x + 5$;   30

4. $y = -\frac{7}{4}x + 11\frac{1}{2}$;   (1) $9\frac{3}{4}$ (2) $6\frac{4}{7}$

5. $y = \frac{3}{8}x + 7\frac{1}{2}$ 6. $y = 0.88x + 0.28$

7.   a) $y = -4.65x + 516$ b) $y = -7x + 743$
8. $y = 400 - 5t^2$ (1) 400 metres   (2) 8.9 secs   (3) 6.3 secs

9. $y = \frac{6}{x} + 4$ a) 4.6 pence   b) 1200 metres.

Exercise B   Page 76
1.   0.94   2.   1   3. -1 4.   0.99   5.   0.89
6. -0.88 7.   0.46   8.   0.14
9.   a)   0.33   b)   0.57   c)   0.60   d)   0.39

Chapter 8
Exercise A   Page 83

1. $\frac{5}{6}$ 2. a) $\frac{3}{13}$ b) $\frac{2}{13}$ 3. a) $\frac{1}{6}$ b) $\frac{1}{3}$ c) 0   d) $\frac{1}{2}$

4. a) $\frac{1}{13}$ b) $\frac{1}{52}$ c) $\frac{1}{4}$ d) $\frac{3}{13}$ e) 0

5. a) $\frac{25}{51}$ b) $\frac{2}{51}$ c)   either $\frac{4}{51}$ or $\frac{3}{51}$ depending on whether the black card removed was a king.

6. $\frac{1}{3}$ 7. $\frac{3}{4}$ 8. a) $\frac{1}{20}$ b) $\frac{85}{99}$

9. a) $\frac{2}{5}$ b) $\frac{3}{5}$ c) 0   d) 1   10. a) $\frac{1}{9}$ b) 0   c) $\frac{1}{12}$ d) 0   e) 1

11. $\frac{1}{4}$ 12. a) $\frac{1}{8}$ b) $\frac{1}{2}$ c) $\frac{1}{8}$ 13. a) $\frac{1}{12}$ b) $\frac{1}{12}$ c) 0

14. a) $\frac{1}{4}$ b) $\frac{1}{2}$ 15. a) $\frac{1}{2}$ b) $\frac{9}{16}$ 16. a) $\frac{1}{4}$ b) $\frac{1}{4}$

17. a) $\frac{1}{8}$ b) $\frac{1}{6}$ c) $\frac{1}{12}$ d) $\frac{1}{4}$ 18. a) $\frac{1}{12}$ b) $\frac{1}{24}$ c) $\frac{1}{2}$

19. $\frac{1}{16}$ 20. $\frac{1}{32}$

Exercise B   Page 85

1. $\frac{1}{64}$ 2. $\frac{1}{36}$ 3. a) $\frac{7}{15}$ b) $\frac{1}{15}$

4. $\frac{7}{24}$ 5. $\frac{1}{663}$ 6. a) $\frac{2}{15}$ b) $\frac{7}{30}$ c) $\frac{49}{575}$

7. a) $\frac{100}{2303}$ b) $\frac{19}{98}$ c) $\frac{1}{2303}$ d) $\frac{969}{46060}$

1.   50   2. *a*) 30   *b*) 60   *c*) 40
3. *a*) 20   *b*) 40   4. 80   5. 20
6. *a*) 46   *b*) 75   7. 200   8. 367
9.   80   10. 9   11. 120

Chapter 9

Exercise A   Page 90

1. 20,   504,   5040,   6720.   n(n − 1)
2. 60   3. 24, $\frac{1}{4}$ 4. 2520

5. 35, $\frac{1}{7}$ 6. 110   7. 151200, $\frac{1}{151200}$

8. 180   9. 604800   10. 59049, $\frac{1}{59049}$

11. 60480   12. 168   13. $\frac{1}{500}$, $\frac{8}{375}$.

Exercise B   Page 91

1. 10,   35,   28,   6,   28,   6.
2. 84   3. 270725   4. 120
5. 220   6. 2156   7. 35

8. 4845   9. 924   10. 77, $\frac{1}{12}$

11. 1260, $\frac{1259}{1260}$ 12. 28, $\frac{1}{7}$

Chapter 10

Exercise A   Page 98

1. *a*) $\frac{80}{243}$ *b*) $\frac{20}{243}$ *c*) $\frac{64}{729}$ 2. $\frac{1}{81}$, $\frac{8}{81}$, $\frac{24}{81}$, $\frac{32}{81}$, $\frac{16}{81}$

3. $\frac{35}{128}$ 4. 0.6561, 0.2916, 0.0486, 0.0036, 0.0001

5. $\frac{7}{64}$ 6. $\frac{16}{81}$, $\frac{32}{81}$, $\frac{24}{81}$, $\frac{8}{81}$, $\frac{1}{81}$ 7. $\frac{459}{512}$ 8. $\frac{44}{125}$

9. *a*) $\frac{2187}{16384}$ *b*) $\frac{2835}{16384}$ 10. 0.432   11. $\frac{53}{3125}$

12. $\frac{23}{648}$ 13. *a*) $\frac{5}{16}$ *b*) $\frac{21}{32}$ 14. *a*) $\frac{280}{2187}$ *b*) $\frac{696}{729}$

15. $\frac{7}{8}$, $\frac{5}{16}$ 16. 0.22, 0.40, 0.53, 0.64

17. *a*) 0.6561   *b*) 0.2916   *c*) 0.0523   *d*) 0.9999

18. $\frac{243}{3125}$, $\frac{810}{3125}$, $\frac{1080}{3125}$, $\frac{720}{3125}$, $\frac{240}{3125}$, $\frac{32}{3125}$ *a*) 5   *b*) 39

19. *a*) 20 *b*) 44   20. *a*) 37 *b*) 90

Chapter 11

Exercise A   Page 107
1. *a*) 1.33   *b*) −1   *c*) 0.273   *d*) −1.6
2. *a*) −0.455 *b*) 1.273   *c*) −2.455   3. She did better in 1st exam.
4.   Jane   5. He did equally well in all three subjects
6.   Bob   7. *a*) May   *b*) Nov.   *c*) Nov. maths   *d*) Nov. English
8. *a*) 34.2 *b*) 37.5 *c*) 25.5 *d*) 27.96   9. 3.85
10.   10   11. 21.5   12. 63.15   13. $\mu = 50$; $\sigma = 5$
14. $\mu = 65$; $\sigma = 10$

Exercise B   Page 110
1.   0.403   2. 0.492   3. 0.226   4. 0.353
5.   0.617   6. 0.953   7. 0.300   8. 0.258
9.   0.089   10. 0.441   11. 0.115   12. 0.001
13.   0.268   14. 0.081   15. 0.983   16. 0.691
17.   0.907   18. 0.633   19. 0.218   20. 0.782
21.   0.903   22. 0.189
23. *a*) 0.614 *b*) 0.242   *c*) 0.092   *d*) 0.159   *e*) 0.309
24. *a*) ± 1.17 *b*) ± 2.20   *c*) −0.46   *d*) 0.87   *e*) −0.24   *f*) 1.28
            g) ±2.46   *h*) ±2.75   *i*) ±2.32 or 2.33

Exercise C   Page 112
1. *a*) 0.023   *b*) 0.159   *c*) 0.081   *d*) 0.023   *e*) 0.954
2. *a*) 0.933   *b*) 0.818

Exercise D   Page 113
1.   0.004   2. 1.56   3. 0.261   4. 11.5%
5.   0.008; 0.044 6. 186.4 cm   7. 0.025   8. 40.8%
9.   77%   10. 9   11. 7 years
12. *a*) 115   *b*) 168.3 cm and 176.7 cm   13. 109.860 g and 160.445 g
14. $\mu = 22.72$; 21.04   15. 47   16. 964
17.   18.87   18. 128.8   19. $\mu = 167.1$ cm $\sigma = 9.43$ cm
20. $\mu = 64.9$; $\sigma = 9.55$

Exercise E   Page 118
1. *a*) 0.029   *b*) 0.08   2. 0.112   3. 0.123
4.   0.097   5. *a*) 0.013   *b*) 0.986   6. 0.006
7.   17   8. 0.999   9. 0.002   10. 0.999
11.   0.031   12. 23

Chapter 12

Exercise A   Page 127
1. *a*) 0.037   *b*) 0.001   2. *a*) 0.084   *b*) 0.510
3.   0.019   4. 0.115   5. From 29.665 to 30.335
6.   £1.95   7. *a*) 96   *b*) 166   8. 240

Exercise B   Page 130
1. 30±0.96   2. 65±0.65   3. 12.6±0.74
4. 171.05±0.865 cm   5. 161.25±1.08 cm 6. 2.5±0.18

Chapter 13

Exercise A   Page 134

1.   No, since prob. = 0.18 for a two-tailed test.
2.   No, since the Z score is 4.95, much higher than the critical value of Z = 1.96 in a two-tailed test.
3.   No, since prob. = 0.075 for a one-tailed test.
4.   Yes, since prob. = 0.007 for a one-tailed test.
5.   No, since prob. = 0.071 for a one-tailed test.
6.   Yes, since prob. = 0.086 for a two-tailed test.
7. *a)*   No, since prob. = 0.133 for a one-tailed test.
   b)   Yes, since prob. = 0.013   *c)* 55 ± 1.76
8.   Yes, since prob. = 0.001 for a one-tailed test.
9.   Yes, since Z is much less than the critical value of -1.64 in a one-tailed test.
10.   Yes, since prob. = 0.006 for a one-tailed test.